江苏大学英文教材基金资助出版

COLLEGE PHYSICS EXPERIMENT

大学物理实验

主编

葛 勇
(Ge Yong)

江苏大学出版社
JIANGSU UNIVERSITY PRESS

镇 江

图书在版编目(CIP)数据

大学物理实验＝College Physics Experiment：英
文 / 葛勇主编. —镇江 ：江苏大学出版社,2021.12
ISBN 978-7-5684-1733-4

Ⅰ.①大… Ⅱ.①葛… Ⅲ.①物理学－实验－高等学
校－教材－英文 Ⅳ.①O4-33

中国版本图书馆 CIP 数据核字(2021)第 260342 号

大学物理实验

College Physics Experiment

主　　编/葛　勇
责任编辑/郑晨晖
出版发行/江苏大学出版社
地　　址/江苏省镇江市梦溪园巷 30 号(邮编：212003)
电　　话/0511-84446464(传真)
网　　址/http://press.ujs.edu.cn
排　　版/镇江文苑制版印刷有限责任公司
印　　刷/广东虎彩云印刷有限公司
开　　本/787 mm×1 092 mm　1/16
印　　张/7.5
字　　数/217 千字
版　　次/2021 年 12 月第 1 版
印　　次/2021 年 12 月第 1 次印刷
书　　号/ISBN 978-7-5684-1733-4
定　　价/38.00 元

如有印装质量问题请与本社营销部联系(电话：0511-84440882)

Preface

In recent years, education for the overseas undergraduates in Jiangsu University has been witnessed a rapid development. The number of the overseas students in the campus increases continuously, and most of these students major in science and engineering.

The college physics experiment is a basic course for the students in science and engineering, which mainly trains the basic experimental skills and improve the practical and innovative abilities of the students. Therefore, this course is also a compulsory course for the overseas undergraduates majoring in science and engineering.

However, up to now, the college physics experiment teaching for overseas students is limited by the lack of a suitable teaching material in English version. To overcome the problem, on the basis of the book of college physics experiment in Chinese and the learning situation of the overseas students in Jiangsu University, we write this book, which has several features as follows:

(1) Low starting point and concise content. Note that the base of the physics for most of the overseas undergraduates are very weak, we simplify the content and reduce the difficulty of this course. Besides, we add the part of preparation experiment, which consists of 4 simple experiments covering the fields of mechanism, electrics and optics. These experiments will help the students to mast the basic experimental instruments as well as the corresponding operations.

(2) Focus on the basics and frontiers of the physics. Most of the experiments in this book correspond to physical laws which are basic and important in physics. For example, in the "Experiments on air cushion track" of Chapter 4, we try to deepen the understanding of the law of conservation of momentum for students by demonstrating the collision between two sliders on the air track. Additionally, in the experiment of "Measurement of magnetic field in electrified solenoid with Hall sensor", students can simultaneously learn the application of the Hall sensor and understand the theory of the Hall effect which can be extended to the frontiers of the physics, such as quantum Hall and quantum spin Hall effect.

(3) Aid to cultivate the innovation ability for undergraduates. Beside the basic

and normal experiments, we also list 4 innovative experiments, in which one or two purpose are proposed without any specific experimental steps. Students can explore and learn the experiment by themselves, and then design each step and finally finish the experiment. By performing these innovative experiments, students can improve their independent learning, comprehensive practice and innovative ability.

There are totally 24 experiments in this book, which originate from the projects of physics experiment designed and developed in the teaching process of the teachers in the collage physics experimental center of Jiangsu University for years. During the process of writing, we gain the strong support and concern from the leaders of overseas collage, physics and electric engineering collage and teachers of the collage physics experimental center. Besides, we are also supported by the Press of Jiangsu University. Here, we express our thanks sincerely.

However, limited by the knowledge level of the authors, there may exist errors and omissions in this book inevitably. We hope that the readers can suggest valuable advices and comments for the further improvement of the book.

Ge Yong
September 2021

Contents

Chapter 1
Measuring error and data processing

The measurement of physical quantity is very important in college physics experiment. However, it is impossible to obtain the true value of the physical quantities owing to the influences and errors from the instruments, measuring method, measurement environment, conditions, operators and so on. Therefore, it is essential to evaluate the reliability of the measured results, estimate the error range, and exhibit the measurement results correctly.

 ## 1.1　Measurement and error

1.1.1　Direct measurement and indirect measurement

People investigate the motion of the objective matters by performing the physical experiment under various experimental conditions, in which measurement is the basis.

In the measurement, we should first select an unit, and then compare the object to be measured with the unit. The ratio between the object to be measured and the unit is the measured value which is highly dependent on the selected unit. Therefore, the correct expression of the measurement results must contains the measured value and unit.

Depend on the method to obtain the measurement results, we can divide the measurement into two types, namely the direct measurement and indirect measurement. In the direct measurement, we compare the object with the selected instruments or other measuring tools, and read the value of the physical quantities directly from the instruments and measuring tools. For example, we usually measure the length, mass and current directly by using meter ruler, balance and ammeter, respectively. However, for indirect measurement, the measured results should be calculated based on the mathematical formulas by using the data read from the instruments or measuring tools. For instance, in the experiment of measuring resistance by voltammetry, the unknown resistance R is calculated by $R = U/I$, in which the voltage U and the current I are measured directly by a voltmeter and ammeter, respectively.

1. 1. 2   Measurement error

The true value is objective size of the physical quantity. In the measurement, we always try to obtain the true value of the physical quantity. However, as aforementioned, owing to the limitation of the instruments, method, environment, measuring conditions, observer and other factors, we usually cannot obtain the true value. There always exists difference between the measuring result and true value. We define the difference as measurement error, which could be expressed as both absolute and relative forms.

The absolute error is a dimension quantity and indicates the degree of the measured result deviating from the true value, which could be expressed as

$$\Delta X = X - A \tag{1-1}$$

Here, X is the measured result, A is the true value, and ΔX is the absolute error.

The relative error is a dimensionless quantity, and usually written as

$$E = \frac{|\Delta X|}{A} \times 100\% \tag{1-2}$$

which indicates the accuracy of the measurement.

As well known that the true value is an ideal concept, it is impossible to capture the true value A in the view of experimental measurement. So we cannot use formula (1-1) or (1-2) to calculate the measurement error. To solve the problem, we use the conventional true value X_0 to replace the true value A. As a result, the relative error turns to be percentage error E, which is expressed as

$$E = \frac{|\Delta X|}{X_0} \times 100\% = \frac{|X - X_0|}{X_0} \times 100\% \tag{1-3}$$

We emphasize that the error always exists and it is impossible to be avoided in the experimental measurement. Therefore, we should study the origin of the error, find a way to reduce the influence of the error firstly, and then try to make the measured value closest to the true value in the same experimental conditions. Finally, we have to evaluate the measurement errors which cannot be eliminated in the experiment.

1. 1. 3   Classification of errors

According to the origin and properties of the error, we can divide the measurement error into three types, named as systematic error, random error and gross error.

1. 1. 3. 1   Systematic error

If the measurement results always deviate from the true value with a fixed value or varies in a certain law during the multi-measurements with the same conditions (e.g., method, instrument, environment, and operator), the error is defined as systematic error.

The origins of the systematic error are listed as follows:

(1) Instrument error. The instrument error is induced by the defect or incorrect operation of the instrument, such as the inaccurate zero point of the spiral micrometer and unequal arms of the balance.

(2) Theoretical error. In physics, some theoretical formula is only valid in the ideal or very critical conditions which are hard to be achieved in experiment. For example, the period of the simple pendulum is usually calculated by using the formula $T = 2\pi\sqrt{\dfrac{l}{g}}$, which is only valid with a tiny swing angle and the influence from the air resistance is ignored. However, a large swing angle and the air resistance is inevitable during the measurement. Therefore, the error induced by the above factors is defined as theoretical error.

(3) Environment error. This error arises from the effects of the external environment, such as temperature, light illumination, pressure, humidity and electromagnetic field. For example, the increase of the temperature will cause the increase of the sound speed in air.

(4) Personal error. The error depends on the physical and psychological characteristics of the operators.

The systematic error will be reduced if we adopt the proper method in the experiment.

1. 1. 3. 2   Random error

The random error is unpredictable in the measurement. It behaves as a random value in the multiple measurements with the same experimental condition, which is attributed to a series of uncertain factors, e.g. the sensory judgment of human, the random variation of the external environment, and internal accidental factors of the instrument.

However, if the times of the measurement is large enough with the same experimental condition, the random error will distribute with the Normal distribution (Guassian distribution). The arithmetical average of the random error will approach to zero when the measurement times increase. Therefore, although the random error cannot be eliminated by improving the measurement method, we can reduce the random error by increase the number of measurement times.

1.1.3.3 Gross error

The gross error is mainly induced by the mistakes occurred in the measurement. If we have rigorous scientific attitude and meticulous working attitude, we can avoid the gross error in the measurement, or find and delete the gross error after the experiment.

1.2 Measurement result and uncertainty

1.2.1 Expression of the measurement result and uncertainty

Generally, the measurement result should include the measured value and the corresponding measurement error in scientific experiments. According to the technical measurement specification of China, the measurement result should be expressed as

$$Y = X \pm \Delta_X \tag{1-4}$$

where Y is the physical quantity to be measured, X is the measured value, and Δ_X is the total uncertainty of the measurement. Note that X and Δ_X have the same unit.

The uncertainty is the evaluation of the measurement error, giving the value range of the measurement in which the true value may locate. The uncertainty is always positive, indicating the degree of uncertainty of the true value to be measured owing to the existence of the measurement error. The reliability of measurement results increases with the decrease of the uncertainty.

Therefore, the true meaning of the formula (1-4) can be read as that the true value will locate in the range $[X - \Delta_X, X + \Delta_X]$ with certain possibility, or there exists a certain possibility for the range $[X - \Delta_X, X + \Delta_X]$ to include the true value. It is worthy to point out that, the range $[X - \Delta_X, X + \Delta_X]$ is a confidence interval, and the possibility is a confidence possibility. The confidence possibility is proportional to the size of the confidence interval, showing an one-to-one correspondence with each other.

To exhibit the measured result more accurately, we also define the relative uncertainty of the measured value, namely

$$E = \frac{\Delta_X}{X} \times 100\% \tag{1-5}$$

Usually, according to the classification of the measurement error, the uncertainty contains A and B components, in which the A component (referred as Δ_A) is evaluated by the statistical distribution of the measured values, and the B component (referred as Δ_B) is evaluated based on the experience or other non-statistical information.

1.2.2　Uncertainty of direct measurement result

1.2.2.1　Uncertainty of the multi-measurement results

In the physics experiment, when a physical quantity is measured many times, we take the arithmetical average value $\overline{X}$ as the measured value, and the total uncertainty is given by

$$\Delta_X = \sqrt{\Delta_A^2 + \Delta_B^2} \tag{1-6}$$

$$\Delta_A = \sqrt{\frac{\sum_{i=1}^{n}(X_i - \overline{X})}{n(n-1)}} \tag{1-7}$$

$$\Delta_B = \Delta_{ins} = \frac{\sigma_{ins}}{\sqrt{3}} \tag{1-8}$$

where n is the number of the measurement times, X_i is the measured value for the i-th measurement, Δ_{ins} is the uncertainty arising from the instrument, and σ_{ins} is the limitation of instrument error. In Table 1-1, we list the errors of several common instruments.

Table 1-1　Errors of several common instruments

Instrument	Specification	Limitation of the error
Vernier caliper	0.02 mm and 0.05 mm	Scale interval
Spiral micrometer	0~25 mm and 25~50 mm	0.004 mm
Balance		Half of the scale interval
Electric meter		Range×Accuracy class%
Digital instrument		minimum displayed value
Resistance box		Reading value×Accuracy class%+Zero resistance
Others		Half of the minimum division value

1.2.2.2　Uncertainty of the single measurement result

In some practical scenes, multi-measurement cannot improve the accuracy of the measurement owing to various limitations, such as the instrument with very low accuracy and the object to be measured is instable. In these situations, we just perform the measurement for only one time. Therefore, $\Delta_A = 0$, and total uncertainty is only determined by Δ_B. It is expressed as

$$\Delta_X = \Delta_B = \Delta_{ins} \tag{1-9}$$

1. 2. 3 Combination of the uncertainty of indirect measurement result

In fact, most of the measured results are calculated based on the physical formulas by using the direct measured values with errors. Therefore, the calculated results will also take the error combined by the errors of the direct measured values. It is so-called "Error propagation".

As mentioned above, we use the uncertainty rather than the measurement error to evaluate the accuracy of the measurement, owing to the impossibility of measuring true value. So, the "Error propagation" becomes the "Propagation of uncertainty" or "Combination of uncertainty". Here, we introduce a method to calculate the uncertainty for the indirect measured results.

Suppose that, x, y, z, $\cdots$ are the direct measured results, N is the physical quantity to be calculated. The relationship between N and x, y, z, $\cdots$ is given by

$$N = f(x, y, z, \cdots) \tag{1-10}$$

The absolute and relative uncertainty of the indirect measured result N could be written as

$$\Delta_N = \sqrt{\left(\frac{\partial f}{\partial x}\right)^2 \Delta_x^2 + \left(\frac{\partial f}{\partial y}\right)^2 \Delta_y^2 + \left(\frac{\partial f}{\partial z}\right)^2 \Delta_z^2 + \cdots} \tag{1-11}$$

$$E = \frac{\Delta_N}{N} = \sqrt{\left(\frac{\partial \ln f}{\partial x}\right)^2 \Delta_x^2 + \left(\frac{\partial \ln f}{\partial y}\right)^2 \Delta_y^2 + \left(\frac{\partial \ln f}{\partial z}\right)^2 \Delta_z^2 + \cdots} \tag{1-12}$$

where the Δ_x, Δ_y and Δ_z are the uncertainties of the direct measured values x, y and z, respectively. This method is called "Square root synthesis".

1. 2. 4 Significant digit

1. 2. 4. 1 Concept of the significant digit

In both direct and indirect measurements, it is very essential to record and process the data accurately. We introduce the concept of the significant digit to exhibit the measured results correctly and effectively.

Figure 1-1 shows the schematic of the measurement of the object's length by using a meter ruler, in which the minimum scale interval is 1 mm. The left end of the object is aligned with the zero line of the meter ruler, the right end of the object locates at the position between 21.7 cm and 21.8 cm, and the final reading could be 21.78 cm. We can see that 21.7 is accurate number read directly from the ruler, and the last estimated number 8 is imprecise. Here, the significant digit of the measured results is written as 21.78 cm. Therefore, it is concluded that the significant digit always consists of several accuracy numbers and one inaccurate digit at the last decimal place.

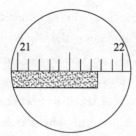

Figure 1-1　Schematic of the measurement with a ruler, unit is mm

1. 2. 4. 2　Significance of significant digit

(1) The place of the inaccurate digit in significant digit reflects the magnitude of the uncertainty. For example, the uncertainty of measured results 15.3 mm is larger than that of 15.79 mm, because the inaccurate digit for 15.3 mm is 0.3 mm but that for 15.79 mm is only 0.09 mm.

(2) We can determine the magnitude of the relative uncertainty according to the number of the significant digit. For instance, the absolute uncertainties of the measured results 2.3 mm and 22.3 mm are located at the same magnitude, however, the relative uncertainty of the 2.3 mm is larger than that of 22.3 mm.

1. 2. 4. 3　Record of significant digit

(1) The inaccurate digit in the significant digit locates at the place of the instrument error. For example, the instrument error for vernier caliper is 0.02 mm (see Table 1-1), the corresponding measured result is 15.46 mm, therefore, the inaccurate digit of the significant digit is 6 locating at the second decimal place which is the same as the instrument error.

(2) The zero at the middle and end of the measured result should be counted as the significant number. For example, the number of the significant digit of 12.050 cm is five, while that of 0.056 cm is two.

(3) When we record the result in the form of scientific expression, the significant digit is the numbers before the power exponent of 10. In addition, the decimal point should locate after the first number of the significant digit.

(4) The last number of the significant digit (inaccurate digit) should locate the place of the uncertainty. Usually, the total uncertainty contains $1 \sim 2$ significant numbers. Here, we agree that the total uncertainty only takes one significant digit.

1. 2. 4. 4　Calculation of significant digit

Owing to the inaccurate digit locating at the last place, the calculation of the significant digit should obey the following rules:

(1) The place of the last number in significant digit is determined by the uncertainty.

(2) For addition and subtraction, the significant digit of the result should be done by rounding up or rounding down the digits after the inaccurate digit. For example, $3\overline{4} + 25.\overline{9} = 59.\overline{9} \approx 6\overline{0}$, in which the inaccurate digits are labeled by overlines, and the final result is obtained by the rounding method and its inaccurate digit aligns with the highest place of the inaccurate number in the addends.

(3) For multiplication and division, the number of significant digits is the same as that with the shortest significant digits among the quantities involved in the calculation. For instance, $1.7\overline{2} \times 4.\overline{1} = 7.\overline{052} \approx 7.\overline{1}$, in which the final result 7.1 has two significant digits which is the same as that of the multiplier 4.1.

(4) For power and root, the number of significant digits is the same as that of the basic digit. For example, $3.\overline{5}^2 = 1\overline{2}.\overline{25} \approx 1\overline{2}$, in which the number of significant digits in both basic digit and result are 2.

(5) For logarithm operation, the decimal place of the significant digits is the same as the number of the antilogarithm. For example, $\ln 267 \approx 5.567$.

(6) During the calculation, the number of the significant digit of the constant, such as $1/4$ and π, is infinite.

1.3 Basic method for data process

Data process is an essential component of the physics experiment, including the data record, reorganization, calculation, analysis and so on. We can obtain the experiment results, reveal the underlying physical phenomena and find the empirical formula through data process. Here, we will introduce four basic methods of data process, including tabulation method, drawing method, successive differential method and the least square method.

1.3.1 Tabulation method

Tabulation method is a basic method to deal with the measurement data. It can exhibit the data methodically and clearly, which is benefited to demonstrate the relationship among various physical quantities. In addition, we can improve the efficiency and reduce or avoid the mistakes of the data process by using tabulation method. The principles of this method are listed as follow:

(1) The name shall be filled in each column of the table, and the unit is added in the headers of the table.

(2) The data recorded in the table should be the original data. Besides, some intermediate results in data processing can also be listed in the table.

(3) The column order of the table should be convenient for calculation, and demonstrate the relation between the recorded data.

(4) The significant digit of the data in the table can express the measurement results correctly.

1.3.2   Drawing method

The drawing method is a common method for data process. We can clearly display the relationship of the measured data together with its variations in the picture. Especially, with this method, we can find the empirical formula for the measured data which is hard to describe by analytic function formula.

According to the different purposes, the drawing method is divided into graphic display method and graphic analysis method.

1.3.2.1   Graphic display method

We can exhibit the relationship of the measured physics quantities by using the graphic display method.

If the relationship between the two physical quantities obeys certain laws, e.g. the linear relationship between the resistance and temperature, we can draw a smooth curve to demonstrate the variation trend of relationship based on the measured data. In this case, it does not require the smooth curve to pass through all the measured points owing to the existence of the measurement error. However, the measured points still need to be distributed equally around the smooth curve, as shown in Figure 1-2.

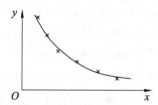

Figure 1-2   Schematic of the graphic display method with smooth curve

If there is no dependency relationship of the two physical quantities, e.g. the variation of temperature with time, we use straight line to connect each neighbor measured points, therefore, the lines pass through all the measured points, as shown in Figure 1-3.

Figure 1-3   Schematic of the graphic display method with broken lines

To show the relationship of the measured quantities with the graphic display method accurately, we should notice the following notes:

(1) We should select the proper coordinate paper, and draw the scale, coordinate axes, symbol of the physical quantity, unit, and scale value.

(2) We should select the proper coordinate origin point to avoid the emergence of the large blank area in the picture. The coordinate grids should match with the minimum division value. The ratio between x and y axis should be set properly to make the curves distribute equally over the most space of the coordinate paper.

(3) The fitting curve (line) should pass through or try to approach to the measured points which marked as the symbols of " + ", " − ", " * " or " ∘ " in the picture.

(4) We should add the detailed captions for each figure, including the name of the figure, experimental condition, author and the date of the figure.

1.3.2.2 Graphic analysis method

We can solve the physical problems quantitatively by using the plotted figure. For example, the relationship between the resistance and temperature could be described by the linear formula $R_t = R_0(1 + bt)$. To determine the coefficient b, we can plot the line of $R_t(t)$ in figure based on the resistance values measured at different temperatures in the experiment, and then find the slope k of the line which equals the coefficient b.

However, in some practical problems, there is no linear relationship between the physical quantities to be measured. In this case, we can convert the nonlinear relationship to the linear one. For example, in the measurement of the gravitational acceleration with a single pendulum, the period of the single pendulum is given by

$$T^2 = \frac{4\pi^2}{g}L \tag{1-13}$$

If we set $y = T^2$ and $x = L$, formula (1-13) is converted to a linear equation $y = \frac{4\pi^2}{g}x$. We can obtain the gravitational acceleration g by plotting the $y(x)$ line and calculating its slope.

1.3.3 Successive differential method

Here we introduce the successive differential method to process a series of measured values in an equal difference sequence.

For instance, we add the weight with mass of m one by one at the lower end of a spring with initial length of x_0, the corresponding lengths of the spring are recorded as x_1, x_2, x_3, x_4, and x_5, respectively. By using the successive differential method, the elongation of the spring e induced by each weight is calculated as

$$e = \frac{1}{3}\left(\frac{x_3 - x_0}{3m} + \frac{x_4 - x_1}{3m} + \frac{x_5 - x_2}{3m}\right) \quad (1\text{-}14)$$

where we should note that $x_3 - x_0$, $x_4 - x_1$ and $x_5 - x_2$ are induced by three weights. This method takes advantages of high utilization of data, small random error and accurate significant figures.

1.3.4　The least square method

Although the graphic display method can exhibit the measurement results intuitively, additional error may be introduced into the final results from the plotted figure owing to the difference between the subjectivity of different authors. To solve the problem, the least square method is adopted to standardize the plotting method.

As aforementioned, the most of the nonlinear functions could be converted to the linear ones, therefore, we only discuss the uni-variate linear fitting problem by using the least square method here for simplify the problem. Suppose that, in the experiment, x_1, x_2, $\cdots$, x_n is a series of controllable physical quantity, and y_1, y_2, $\cdots$, y_n are the corresponding measured values. The relationship between x and y is linear. Commonly, we just need to select two groups of measured data from (x_i, y_i) to plot a straight line, but this line may have a large error. The least square method provides a mathematical analysis way to find a fitting line with the minimum error for all the measured data. This fitting line will not have to pass through all the data points, it just passes by these points in the closest way, as shown in Figure 1-4.

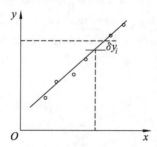

Figure 1-4　Fitting line plotted by the least square method

If the function of the fitting line in Figure 1-4 is set as $y = a + bx$, according to the least square method, the coefficients a and b will be given by

$$b = \frac{\overline{xy} - \overline{x} \cdot \overline{y}}{\overline{x^2} - (\overline{x})^2} \quad (1\text{-}15)$$

$$a = \overline{y} - b\,\overline{x} \quad (1\text{-}16)$$

where $\bar{x} = \frac{1}{n}\sum x_i$, $\bar{y} = \frac{1}{n}\sum y_i$, $\overline{xy} = \frac{1}{n}\sum (x_i y_i)$, and $\overline{x^2} = \frac{1}{n}\sum x_i^2$. Based on the formulas (1-15) and (1-16), we can obtain the function of the fitting line for any series of measured data.

1.4 Introduction of basic instruments for physical experiments

Length, mass and time are three fundamental physical quantities in mechanics. The instruments, utilized to measure the above three quantities, play the prime role in mechanical or even whole physical experiment region.

1.4.1 Measurement of the length

The meter ruler is the most common tool to measure the length. However, the scale value of the meter ruler is only 1 mm, which can not satisfy the measurement requiring high accuracy. To improve the measurement accuracy, various schemes are proposed to design the tools with high accuracy, among which the vernier caliper and screw micrometer are two kinds of the most common representatives.

1.4.1.1 Vernier caliper

Figure 1-5 shows the schematic of the vernier caliper, in which the scale line of the main ruler is the same as that of the meter ruler, and the vernier E can move along the main ruler D. We usually measure the thickness and outer diameter of the object between the A and B ends, measure the inner diameter with A' and B' ends, and measure the depth of the groove with C end. The length of the object is measured by reading the distance between the zero lines on the main ruler and vernier.

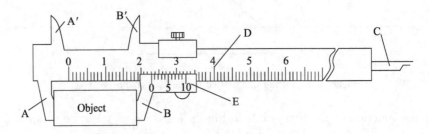

Figure 1-5   Schematic of the vernier caliper

The scale on the main ruler is the same as that on meter ruler, the minimum scale value is 1 mm. Different with that on the main ruler, the length L of n grids on the vernier equals that of $(n-1)$ grids on the main ruler. Therefore, the length difference for each grid on the main ruler and vernier is expressed as $\delta=l-l'$, where $l'=L/n$, $l=L/(n-1)$, and δ is the scale interval of the vernier caliper. For the twenty-scale interval vernier caliper $(n=20)$, the length of the 20 grids on vernier equals 19 mm on the main ruler, therefore, each grid on vernier is about 0.95 mm, and the scale interval of this vernier caliper is 0.05 mm.

Normally, the zero line on the vernier doesn't align with that on the main ruler. As shown in Figure 1-6, the zero line on the vernier locates beyond the 20 mm scale line on the main ruler, and the 3-rd line on the vernier aligns with scale line on the main ruler which indicates that the reading from the vernier is 0.05 mm$\times$3$=$0.15 mm. Therefore, the measured value is 20$+$0.15$=$20.15 mm. It is concluded that the measured value X from the vernier caliper could be obtained by the formula $X=Y+K\delta$, in which Y is the reading determined by the scale line on main ruler at the left side of the zero line on vernier, K indicates the K-th scale line on vernier aligning with that on the main ruler, and δ is the scale interval of the vernier caliper. To make the data reading more conveniently, we label the long scale lines on the vernier with "0, 25, 50, 75 and 100", which represent 0, 0.25, 0.75 and 1 mm, respectively.

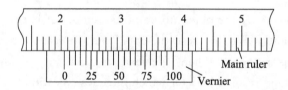

Figure 1-6   Schematic of twenty-scale interval vernier caliper

The notes for the usage of the vernier caliper are listed as follows:

(1) Before the measurement, we should check whether the zero lines on the main ruler and vernier align with each other. If not, please record the error between the two zero lines.

(2) When we read the data from the vernier caliper, we should find and judge which scale line on the vernier aligns with that on the main ruler, which will generate the measurement error.

1. 4. 1. 2   Screw micrometer

Screw micrometer is another kind of tool with higher measurement accuracy than that of the vernier caliper, by which the accuracy of measuring length can reach 0.01 mm. However, its measuring range only covers several centimeters which is obvious shorter than that of vernier caliper. Figure 1-7 shows the structure of the screw micrometer. A is the U-shaped support, in which the end F is fixed, and the other end B is a screw arbor with the pitch of 0.5 mm. The end B will move a distance of 0.5 mm when the screw arbor rotate a full turn. E (main ruler) is a casing pipe on which each small grid indicates 0.5 mm. D (auxiliary ruler) is connected with the screw arbor B, and there are 50 grids around its circumferential direction. The end B will move a distance of 0.01 mm if the D rotates one small grid. The minimum scale interval of the screw micrometer is 0.01 mm. We can also estimate a digit at the third decimal place (0.001 mm) with the unit of mm.

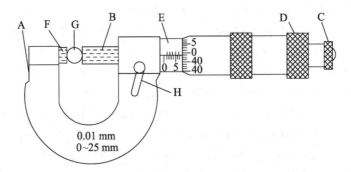

A—U-shaped support; B—Movable end; C—Hand wheel; D—Auxiliary ruler;
E—main ruler; F—Fixed end; G—Object to be measured; H—Locking device.

Figure 1-7   Schematic of screw micrometer

In the measurement, the object to be measured is put between the ends F and B, and clamped tightly. We first read the data from the scale lines on the main ruler E, and record it as x (mm). Then, we read the data from the scale lines on the auxiliary ruler D, and recorded it as n which includes an estimated digit. The length of the object is given by $L = \left(x + \dfrac{n}{100} \right)$.

The notes for the usage of the screw micrometer are listed as follows:

(1) We should note that H in Figure 1-7 is the locking device. Before the measurement, we should turn on the micrometer by rotating the device H.

(2) Before the measurement, we should check that whether the reading is zero when the ends F and B touch with each other tightly. If the reading is non zero, we should use the measured value subtract the non-zero reading.

(3) In the measurement, we should rotate the hand wheel C slowly until we hear the sound "zha" three times. This step is adopted to improve the accuracy of the measurement and avoid to damage the instrument at the same time.

(4) When we finish the measurement, we should keep a small interval between the ends F and B, and do not forget to lock the micrometer.

1.4.2　Measurement of mass

1.4.2.1　Physical balance

Usually, the mass of an object is measured by a balance. In physical experiment, we use the physical balance. The weighing and sensing are two specification parameters of the balance. The weighing indicates the maximum mass that the balance can measure, and the sensing is defined as m/n, in which m is the mass added on the plate of the balance and n is the number of the deviated grids of the balance's pointer shifted away from the zero line. The unit of the sensing is mg/grid.

The main operation steps of the balance are listed below:

(1) Adjust the screws at the bottom, make the base plate of the balance become horizontal.

(2) Move the vernier on the beam of the balance and place it at the zero line. Put up the beam by rotating the brake knob, and then adjust the balance nuts until the pointer of the balance aligns with the zero line.

(3) The object to be measured should be placed on the left plate, and the weights are put on the right plate. We make the balance reach the equilibrium state by adding or removing weights and adjusting the position of the vernier.

(4) When we finish the measurement, we should lower down the beam of the balance.

1.4.2.2　Electric balance

The accuracy of the electric balance could be selected base on the requirement of the measurements. The accuracy of the best electric balance can be higher than that of mechanical analytical balance which is usually utilized in chemistry experiment. The electric balance can measure the mass without any weights, and we can read the data directly from the display screen of the balance. However, it needs to be recalibrated after long term use.

1.4.3　Measurement of time

1.4.3.1　Mechanical stopwatch

The mechanical stopwatch has two pointers, in which the longer one is the minute hand and the shorter one is the second hand. The numbers on the surface of the stopwatch represent the values of seconds and minutes. The scale interval of the stopwatch is 0.2 s or 0.1 s.

On the top of the watch, there exists a button. By pressing it, we can wind and control the working state of the stopwatch. The stopwatch will start timing by pressing the button, and stop timing by pressing the button again. When we press the button third times, the minute and second hands of the stopwatch will return to the zero line.

Before the measurement, we should check the zero state of the stopwatch, record the initial reading value. All the measured values should be substracted the initial reading value. After the measurement, we should let the stopwatch continue to work until its spring relaxes completely.

1.4.3.2　Electric stopwatch

The electric stopwatch has a liquid-crystal screen, on which we can read the measured time directly. The minimum display time is 0.01 s, which is also the instrument error of the stopwatch. The operation steps are similar with the mechanical one that is introduced above.

1.4.3.3　Digital millisecond meter

The digital millisecond meter works based on a high-frequency quartz crystal oscillator which can generate standard time base signals continuously. In experiment, digital millisecond meter times by comparing the standard time base signals with the detected signals input by various photoelectric elements (sensors).

Chapter 2
Basic physical experimental methods

2.1 Common physical experiment ideas

The idea, method and technology of physics experiment are the basis of the knowledge structure of modern high-tech talents, which are also the basis and source of applied technology. Physics experiment plays an important role in the competence education. The experimental course can cultivate rigorous scientific thinking and innovative spirit of the students, and improve their ability to integrate theory with practice. In the process of exploring the unknown world, physics shows a series of scientific world outlook and methodology, which has a profound impact on the basic understanding of the material, human thinking mode and social life. The history of the development of physics has proved that the correct scientific thought as well as the scientific method are the soul of scientific research.

The method of physical experiment is the dynamic interaction between experience (in the form of experiment and observation) and thinking (in the form of creatively constructed theories and hypotheses). Galileo is one of the founders of modern science. He established two research principles for natural science, namely observation experiment and quantitative method, which have expanded the scientific method of combining experiment with mathematics and practical experiment with ideal experiment. The experimental physics created by Galileo, including the design idea and experimental method of the experiment, lead a new way in the development of natural science. In hundreds of years, many outstanding experiments have emerged, as the milestones in the history of physics, showing the extremely rich and wonderful physical ideas as well as their ingenious physical ideas and unique methods of dealing with and solving problems, such as carefully designed instruments, perfect experimental arrangements, superb measurement technology, carefully processing of experimental data and impeccable analysis and judgment, and creative ways and methods to solve problems. These ideas and methods have gone beyond the specific experiment, and have the universal guiding significance in the field of the physical experiment. Learning and mastering the design ideas, measurement and analysis methods of physical experiment will be great benefit to the research of the course of physical experiment and even other disciplines.

2. 2 Basic measurement methods of physical experiments

All physical quantities describing the states and motions of the matter can be derived from several basic physical quantities. The descriptions of these basic physical quantities are only obtained by the physical measurement. With the development of the science and technology, the method and accuracy of the measurement are continuously improved. The measured methods for the same physical quantity in different measurement ranges will be different. Even in the same measurement range, measurement methods are still distinct for different accuracy requirements. The measurement method is determined by the range of physical quantities and the requirements of the measurement accuracy. There are many measurement methods in physical experiments. Here, we only briefly introduce several basic measurement methods for the international students in science and engineering.

2. 2. 1 Comparison method

Comparison method is one of the most basic and important measurement methods. In this method, we measure the physical quantity by comparing it with the similar physical quantity based on the known standard value directly or indirectly. The obtained ratio is the measured value of quantity to be measured. We can divide the comparison method into direct and indirect comparison method according to whether the measured results are converted.

2. 2. 1. 1 Direct comparison method

The simplest direct comparison method is to compare the object to be measured with the standard quantity on the measuring tool directly. For example, we measure the length of an object by comparing it with the standard ruler directly, in which the minimum scale interval millimeter is used as the unit. The direct comparison method has the following characteristics:

(1) The dimension of the measured quantity is the same as that of the standard quantity.

(2) The comparison between the measured quantity and the standard quantity occurs at the same time.

When the difference between the quantity to be measured and the known standard quantity is very tiny, we adopt the coincidence comparison method. In this method, the measurement is extended to several periods until the two quantities coincide with each other. Therefore, the size of the unknown quantity can be obtained through the comparison. For example, the measurement of the length by utilizing the aforementioned vernier caliper adopts the coincidence comparison method.

2.2.1.2   Indirect comparison method

However, some physical quantities are difficult to be measured by direct comparison method. Therefore, the indirect comparison method is adopted. For example, if we want to read the unknown resistance in a circuit, first read the current value I, and then use an adjustable standard resistance to instead the unknown resistance in the circuit. We adjust the value of the standard resistance and keep the voltage of the regulated power supply unchanged, until the current in the circuit becomes I again. In this situation, the value of the standard resistance R_s equals that of the unknown resistance R_x.

2.2.2   Equilibrium method and compensation method

The equilibrium principle is an important basic principle in physics, and the corresponding equilibrium method is an important method to analyze and solve physical problems. In the equilibrium condition, many complex physical phenomena can be described easily, and even some complex physical functions can also be simplified and well understood. The experiment can maintain the original conditions, and the observation has high resolution and sensitivity, therefore, it is easy to realize qualitative and quantitative physical analysis. For example, balance and electronic scale are designed according to the principle of mechanical balance, which are used to measure physical quantities such as mass and density of substance. The bridge circuit, designed on the basis of the equilibrium between electrical quantities such as current and voltage, can be used to measure the electromagnetic characteristic parameters of substance, such as resistance, inductance, capacitance, dielectric constant and permeability.

We can select or adjust the standard value S, and make it equal to that of the physical quantity to be measured (denoted as X), which could offset (or compensate) the effect of the to be measured quantity. In this situation, the system reaches an equilibrium state (or compensated state), and X could be obtained based on the definite relationship between it and the standard value S. This measurement method is called compensation method.

Usually, the compensation method is used together with equilibrium method and comparison method. We measure the quantities by using the equilibrium method, and correct the error using the compensation method. The measurement system of the compensation method includes a standard measure and a balance (or zero indicator). During the measurement, X could be directly compared with the standard quantity S which will be adjusted to make the difference between S and X become zero. The key step of the measurement of compensation method is to make two quantities reach an equilibrium state. This method takes the advantage that some additional system errors can be avoided. If the system has high precision standard measure and balance indicator, it can obtain high resolution, sensitivity and measurement accuracy.

2. 2. 3　Amplification method

In some practical scenes, the physical quantities to be measured are very tiny, such as small length, short time and weak current, which inevitably effect the measurement accuracy by using the conventional measurement method or even make the measurement impossible. To overcome this problem, we usually adopt the amplification method to enlarge the measured quantities before the measurement. Therefore, this method is also a basic measurement method. It is worthy to note that the reduction is also regarded as a kind of amplification whose magnification is less than 1. The amplification method commonly could be divided into accumulated, mechanical, optical, electronic amplification and so on.

2. 2. 3. 1　Accumulated amplification method

In the measurement of gravitational acceleration by using a single pendulum, the period of single pendulum is 2 s, and the instrument error of the mechanical stopwatch is 0.1 s. If we only measure a single period by using this stopwatch, the relative uncertainty of measurement is $0.1/2 = 0.05 = 5\%$. However, if we measure 50 periods of the single pendulum by using the same stopwatch, the relative uncertainty will be $0.1/(2 \times 50) = 0.001 = 0.1\%$. The results indicate that we can reduce the measurement error by using the accumulated amplification method.

2. 2. 3. 2　Mechanical amplification method

Mechanical amplification method is the most intuitive amplification method. For example, the subdivision degree of the measurement tools can be improved by using a vernier. By introducing a vernier with equally distributed n grids, the scale interval of the vernier caliper becomes Y/n, where Y is the original scale value of the main ruler. The mechanical amplification method is also utilized in the screw micrometer, in which the pitch is enlarged by the circumference of the screw nut.

2.2.3.3   Optical amplification method

In optical amplification method, we commonly enlarge the physical quantities by using the optical systems, and then measure the enlarged values and calculate the final result. As shown in Figure 2-1a, if we want to measure the tiny flare angle α of the laser beam emitted from the point C, we should know the length of AB and BC, and calculate the angle by $\tan\alpha = AB/BC$. However, the AB is also very tiny, which will induce a large measurement error. If we measure the corresponding $A'B'$ and $B'C$, the measurement error will be greatly reduced with the same measuring tools. Therefore, the optical amplification method is often referred to as the elongation method.

If the angle α is too small to enlarge $A'B'$ by using the method illustrated in Figure 2-1a, we can increase the optical path by setting two parallel plane mirrors as shown in Figure 2-1b, in which the laser beam undergoes multiple reflections between the two mirrors.

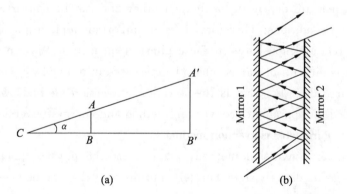

(a)   (b)

Figure 2-1   Schematic of optical amplification method

In the measurement of Young's modulus of the elastic materials, an optical lever is designed to measure the small elongation of wire length, which is a typical example of the optical amplification method.

2.2.3.4   Electronic amplification method

Weak electrical signals can be observed after amplification by amplifiers. If the quantity to be measured is a nonelectrical quantity, it could be converted into the electrical quantity by sensors, and then enlarged by amplifiers. This electronic amplification method is widely used in electromagnetic measurement. The amplification of electrical signals contains voltage amplification, current amplification and power amplification, in which the electrical signals could be AC or DC. With the development of the microelectronic technology and electronic devices, it is very easy to realize the amplification of various electrical signals. Therefore, electronic amplification method is the most widely used and common measurement method. For example, triode is a common component that can be encountered in any electronic circuit, because any small change of the grid voltage U will induce a large collecting plate current I. Therefore, the triode is often used as amplifiers. Recently, with the emerging of a variety of new highly integrated operational amplifiers, it is very easy to amplify the weak current signals by several to more than ten orders of magnitude. Therefore, in physical experiments, the nonelectrical quantities are often converted into the electrical signals which are amplified and then converted back (such as piezoelectric conversion, photoelectric conversion, electromagnetic conversion, etc.). When we amplify the electrical quantity, we should reduce the background signal, improving the signal-to-noise ratio of the measurement.

2. 2. 4　Conversion method

In physics, there always exist intrinsic relationships between different quantities, which are dependent with each other and can be converted into each other in certain conditions. Therefore, it is an important topic in physics to study the intrinsic relationships between these physical quantities. We can convert the quantity that is difficult or impossible to be measured into the physics quantity that can be measured directly. That is the so-called conversion method, which can be roughly divided into parameter conversion method and energy conversion method.

2. 2. 4. 1　Parameter conversion method

In parameter conversion method, we convert the physics quantity that is difficult to measure directly and accurately into the ones that can be measured easily and accurately based on the transformation relationships or laws between these parameters. A classic example is the measurement of the volume and density of the irregular objects firstly performed by Archimedes, who is a famous scientist in ancient Greece. Owing to the irregular shape of the object, we cannot measure its volume directly. An approach to overcome this problem is proposed by Archimedes. We can first measure the mass m of the object in the air, and then immerse the object into a water with density ρ_0, and measure its mass m_1 again. The density of the object is expressed as

$$\rho = \frac{m}{m - m_1} \rho_0 \qquad (2\text{-}1)$$

Therefore, the volume measurement of the object is transformed into the measurement of mass m and m_1.

2. 2. 4. 2   Energy conversion method

In energy conversion method, we convert the physical quantity into the other ones by energy converter. The emergence of new functional materials, such as thermal sensitive, photo sensitive, pressure sensitive, gas sensitive and humidity sensitive materials, provide new schemes to design various sensitive devices and sensors, provide good conditions for the improvement of the scientific experiments and physical property measurements. Usually, we convert the physical quantity to be measured into electrical parameters by using various sensors and sensitive devices, because the electrical instruments are universal and easy to be manufactured, and the measurement of electrical parameters are convenient and rapid. Several common energy conversion measurement methods are listed as follows.

(1) Thermoelectric exchange measurement. In this method, the thermal quantity is converted into electrical quantity. For example, we can measure the temperature based on the principle of thermoelectric electromotive force, in which the measurement of temperature is converted into that of thermoelectromotive force.

(2) Piezoelectric exchange measurement. In this method, we transform the pressure signals into the electric ones by using a transducer fabricated by piezoelectric ceramics, and the reversed process is also permitted. A well-known example is the microphone and loudspeaker.

(3) Photoelectric exchange measurement. In this method, a transducer is utilized to convert the luminous flux into the electricity based on the photoelectric effect. In fact, various photoelectric converters have been widely used in measurement and control systems.

(4) Magnetoelectric exchange measurement. In this method, we exchange the magnetic and electrical quantities by using the semiconductor sensor based on Hall effect.

It could be found that the above energy conversion methods provide a convenient and rapid way to measure the non-electric quantities, such as the displacement and pressure. The various sensors are the key devices of the energy conversion method. In principle, all physical quantities can be converted into other signals for measurement by using the corresponding sensors.

2. 2. 5   Light interference and diffraction method

In the measurement requiring high accuracy, light interference and diffraction are often adopted. In the interference phenomenon, the variation of the optical path difference for adjacent interference fringes equals to the wavelength. Although the wavelength of visible light is too small to be measured directly, the interval of adjacent interference fringes or the number of the interference fringes can be measured. Therefore, it is feasible to obtain the optical path difference by measuring the number of fringes or the change of fringes. For example, we can accurately measure the small length or angle change, small deformation and other related physical quantities, or test the flatness, sphericity, finish of the object surface and the distribution of internal stress in the workpiece based on the equal thickness interference phenomenon of light.

The diffraction method of light can be widely used to measure the size of small objects. It plays an important role in modern physical experimental methods, including spectral technology and methods, X-ray diffraction technology and methods, electron microscopy technology and methods.

Chapter 3
Preparation experiment

 Experiment 1   Measurement of density of solid

I. Purpose

(1) To master the measurement of density of the regular and irregular solid materials.

(2) To further study the usage of the vernier caliper, screw micrometer and physical balance.

II. Instruments

Vernier caliper, screw micrometer, physical balance, object to be measured, measuring cup, thermometer, and so on.

III. Principle

To obtain the density of the object, we should know the volume and mass of the object, which involve the measurement of length and mass. The vernier caliper and screw micrometer can be used to measure the length, and the mass can be measured by the balance. The referred instruments are the basic instruments of physical experiments, and we have already introduced them in Chapter 1. Here, we focus on the measurement of density of regular and irregular solids by using these basic instruments.

1. The formula to calculate the density

The density ρ of an object with mass m and volume V is given by

$$\rho = \frac{m}{V} \tag{3-1}$$

For the regular object, the mass of the object can be measured directly by physical balance, and the length and diameter of the object can be measured by vernier caliper, screw micrometer and other length measuring instruments. In this case, the volume of the object can be obtained by formula (3-1).

2. Measure the density of solids by hydrostatic weighing method

The hydrostatic weighing method can be used to accurately measure the volume of the irregular object, whose volume cannot be obtained from its geometric parameters such as length, width, height, diameter, and so on.

If the volume of the liquid (such as pure water) is the same as that of the object, and the mass of the liquid is m_0, then the volume of the object is

$$V = \frac{m_0}{\rho_w} \tag{3-2}$$

where ρ_w is the density of the liquid. According to formula (3-1) and (3-2), we obtain that the density of the solid object is

$$\rho = \frac{m}{m_0}\rho_w \tag{3-3}$$

Therefore, the measurement of the density of an object is converted into the measurement of the mass. If the mass of the object weighed in air is m, and that measured in water is m_1, according to the principle of Archimedes, the mass of liquid with the same volume is $m_0 = m - m_1$. Substituted it to formula (3-3), the density of the object is expressed as

$$\rho = \frac{m}{m - m_1}\rho_w \tag{3-4}$$

However, in some practical situations, the density of the object is smaller than that of liquid, therefore, we cannot use formula (3-4) to measure the density. To solve the problem, it is necessary to hang a heavy object under the object to be measured. Firstly, we measure the total mass m_3 by keeping the object to be measured above the liquid but the heavy object under the liquid, and then weigh the total mass m_4 by immersing all the object in the liquid. In this situation, the buoyancy of the object in the liquid is $F = (m_3 - m_4)g$, and the density of the object is

$$\rho = \frac{m}{m_3 - m_4}\rho_w \tag{3-5}$$

IV. Contents and steps

1. Measure the density of a hollow cylinder

(1) The inner diameter, outer diameter and height of the hollow cylinder are measured by a vernier caliper, and record as D_1, D_2 and H. Each quantity should be measured six times in different directions, then we should calculate the average values as well as the uncertainties of the above quantities.

(2) The mass m of the hollow cylinder is measured by a physical balance. The operation should follow the relevant contents and requirements of the balance. The instrument error of the balance σ_m is the half of its sensitivity.

(3) Calculate the density of the hollow cylinder based on formula (3-1).

2. Measure the density of a metal sphere

(1) The diameter of the metal sphere D_3 is measured by the screw micrometer, and we should repeat the measurement for six times in different directions. Record the zero value of the screw micrometer d_0.

(2) The mass of the metal sphere m also is measured by the balance.

(3) Calculate the density of the metal sphere and the corresponding uncertainty.

3. Measure the density of a metal nut with hydrostatic weighing method

(1) Hang a metal nut on the left hook of the balance with a thin line and measure its mass m.

(2) Put the glass with water on the plate of the balance, and immerse the metal nut into the water slowly. Drive the bubbles attached on surface of the object by using a glass rod, and then measure the mass m_1 of the metal nut in water.

(3) Measure the temperature of the water with a thermometer, and find out the density ρ_w of the water at the measured temperature.

(4) Calculate the density of the metal nut based on formula (3-3), and calculate the corresponding uncertainty.

(5) Notes for the experiment:

i. When the metal nut is suspended in the water for weighing, we should keep it away from the wall or bottom of the glass, and keep it totally under the water.

ii. The mass of the suspension wire in the experiment is ignored, so the wire should be very thin.

4. Measure the density of paraffin wax

(1) Measure the mass m_2 of the paraffin wax by the balance.

(2) Hang a heavy object under the paraffin wax, place the heavy object under the water and the paraffin wax in air. The mass of the system is measured as m_3.

(3) Put both the heavy object and the paraffin wax totally under the water, and measure the mass m_4 of the corresponding system with a balance.

(4) Calculate the density of the paraffin wax by using formula (3-5), and calculate the corresponding uncertainty.

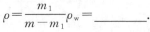

V. Data record and process

1. The density of a hollow cylinder

Scale interval of the vernier caliper = _____ mm,

zero value d_0 = _____ mm, instrument error σ_m = _____ mm,

mass of the hollow cylinder m_1 = _____.

The measured data for the cylinder is recorded in Table 3-1.

Table 3-1 Data record for diameters and height of the hollow cylinder

Measuring times	1	2	3	4	5	6	Average
D_{1i}/mm							
$(D_{1i}-\overline{D_1})^2$/mm²							—
D_{2i}/mm							
$(D_{2i}-\overline{D_2})^2$/mm²							—
H_i/mm							
$(H_i-\overline{H})^2$/mm²							—

$$\rho = \frac{4m_1}{\pi(D_2^2-D_1^2)H} = \underline{\qquad}.$$

2. The density of a metal sphere

Scale interval of the screw micrometer = _____ mm,

zero value d_0 = _____ mm, instrument error σ_m = _____ mm,

mass of the metal sphere m_2 = _____.

The measured values of the diameter is recorded in Table 3-2.

Table 3-2 Diameter of the metal sphere

Measuring times	1	2	3	4	5	6	Average
Diameter d_{3i}/mm							
$(d_{3i}-\overline{d_3})^2$/mm²							—

$$\rho = \frac{6m_2}{\pi D_3^3} = \underline{\qquad}.$$

3. The density of a metal nut with hydrostatic weighing method

Mass of the metal nut in air m = _____,

mass of the metal nut in water m_1 = _____,

temperature of the water t = _____ ℃, density of the water ρ_w = _____,

$$\rho = \frac{m_1}{m-m_1}\rho_w = \underline{\qquad}.$$

4. The density of paraffin wax

Mass of the paraffin wax in air $m_2 =$ _____ ,

mass of the system with the paraffin wax in air and heavy object in water $m_3 =$ _____ ,

mass of the system with both the paraffin wax and heavy object in water $m_4 =$ _____ ,

temperature of the water $t =$ _____ ℃ , density of the water $\rho_w =$ _____ ,

$$\rho = \frac{m_2}{m_3 - m_4} \rho_w = \underline{\qquad} .$$

VI. Questions

(1) When we measure the density of irregular solids, if bubbles are adsorbed on the surface of the object immersed in water, will the density value obtained from the experiment be larger or smaller? Why?

(2) How to use the hydrostatic weighing method to measure the density of the liquid? Describe the principle and steps briefly.

Experiment 2　Measurement of gravitational acceleration

I. Purpose

(1) To master the principle and method of the measurement of gravitational acceleration by using a single pendulum.

(2) To learn the measurement of length and time, and the corresponding data process method.

II. Instruments

Single pendulum, band tape and stopwatch.

III. Principle

Gravitational acceleration is a physical quantity which widely used in mechanics. The gravitational acceleration will vary slightly with the change of geographical position on the earth. In this experiment, we will introduce one method to measure the gravitational acceleration g with a single pendulum.

As shown in Figure 3-1, the single pendulum consists of a suspended line and a small ball, in which the thin line is non-stretchable. If the swing angle is very small (less than 5°), the swing motion of the ball will approach to the simple harmonic vibration with the period of $T = 2\pi\sqrt{\dfrac{L}{g}}$, where L is the length of the pendulum, g is the gravitational acceleration. According to the above formula, we obtain that

$$g = \frac{4\pi^2 L}{T^2} \tag{3-6}$$

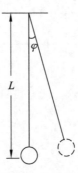

Figure 3-1　Schematic of single pendulum

Therefore, the gravitational acceleration could be calculated by formula (3-6).

In practical measurement, we increase the measurement times (about 50 times of the period) to reduce the error, therefore, the measured time $t=50T$, then the gravitational acceleration is calculated by

$$g = \frac{10^4 \pi^2 L}{t^2} \tag{3-7}$$

IV. Contents and steps

(1) The pendulum length L is measured by a meter ruler.

(2) Adjust the arc ruler. When the ball stops the vibration and becomes static, it will locate at the zero point of the arc ruler.

(3) Place the small ball at a small angle (less than 5°) away from the zero point of the arc ruler and release it, the ball will swing freely. We start timing when the ball swings to the lowest position, and stop timing after 50 periods.

(4) Repeat step 3 for six times.

V. Data record and process

Length of the pendulum $L=$ _____ cm, $g = \dfrac{10^4 \pi^2 L}{\bar{t}^2} =$ _____ .

Record the measured time t_i in Table 3-3, and finish the calculations.

Table 3-3　Data record for the measurement of period and corresponding uncertainty

Measuring times	1	2	3	4	5	6	Average
t_i/s							
$(t_i-\bar{t})^2/\mathrm{s}^2$							—

Experiment 3   Study of the volt-ampere characteristic

I. Purpose

(1) To master the two circuits of the voltammetry, and know their application conditions.

(2) To master the volt ampere characteristic curve of crystal diode measured by volt ampere method.

(3) To learn the usage of the ammeter and voltage divider correctly.

II. Instruments

C_{31}-A typed ammeter, C_{31}-V typed voltmeter, DC regulated power supply, C_{31}-mA ammeter, 2AP-typed crystal diode, sliding wire rheostat, high resistance R_{xh} and low resistance R_{xl} to be measured.

III. Principle

1. Volt ampere characteristic curve

According to Ohm's law, the volt ampere characteristic curve induced by a linear conductor is a straight line passing through the origin point of the coordinate. The reciprocal of the slope of the straight line equals the resistance of the conductor, which is a constant independent of current and voltage.

The volt ampere characteristic curve of an electric element that does not obey Ohm's law is not a straight line. Although the nonlinear element does not obey Ohm's law, its resistance can still be defined by the formula $R = U/I$, which is not a constant but a variable related to the current and voltage.

According to the different behaviors of the volt ampere characteristics, we can distinguish the conductive characteristics and functions in the circuit.

2. Two circuits of voltammetry and their application conditions

The methods to measure the electric quantities by ammeter and voltmeter directly or indirectly are so-called voltammetry. Here, we mainly discuss how to measure the resistance of conductor with voltammetry, in which the resistance is obtained by

$$R = \frac{U}{I} \qquad (3\text{-}8)$$

where U is the voltage at the two ends of the resistance, and I is the current passing through it. Figure 3-2 shows the two circuits for the voltammetry.

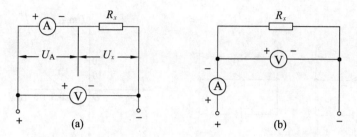

Figure 3-2   Circuits for measuring resistance by voltammetry

As shown in Figure 3-2a, the ammeter is located inside the voltmeter which is called ammeter internal connection method. The current I measured by the ammeter is exactly the same as that passing through R_x. But the voltage U measured by the voltmeter is the sum of the voltage U_x and U_A. According to formula (3-8), the measured value is larger than R_x. If R_A is known, the accurate calculation for R_x is

$$R_x = \frac{U-U_A}{I} = \frac{U}{I} - R_A \tag{3-9}$$

As shown in Figure 3-2b, the ammeter is connected outside the voltmeter which is called ammeter external connection method. The current I measured by the ammeter is the sum of the currents passing through R_x and voltmeter. The voltage U measured by the voltmeter is the same as that at the two ends of R_x. According to formula (3-8), the measured value is smaller than R_x. If R_V is known, the accurate calculation for R_x is

$$R_x = \frac{U_x}{I-I_V} = \frac{U_x}{I\left(1-\frac{I_V}{I}\right)} = \frac{U_x}{I}\left(1+\frac{I_V}{I}\right) = R\left(1+\frac{R}{R_V}\right) \tag{3-10}$$

Generally, due to circuit reasons, the resistance value to be measured is always too large or too small, indicating that there is a certain systematic error. In order to reduce the error, it is necessary to make a rough estimation according to the values of R_x, R_V (resistance of voltmeter) and R_A (resistance of ammeter), and select a suitable measurement circuit.

When $R_x \gg R_A$ and R_V approaches to R_x, we can adopt the circuit in Figure 3-2a to measure the R_x. However, if $R_V \gg R_x$ and R_x approaches to R_A, we can use the circuit in Figure 3-2b to measure the unknown resistance.

3. Volt ampere characteristic of crystal diode

Unidirectional conductivity is the main characteristic of the crystal diode. The 2AP-typed crystal diode is a classical semiconductor made of P-type and N-type germanium. Figure 3-3a and 3-3b show the schematics of the P-N junction with positive and negative voltages, respectively. As shown by Figure 3-3c, the positive

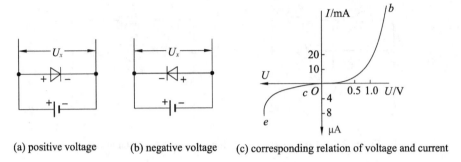

(a) positive voltage   (b) negative voltage   (c) corresponding relation of voltage and current

Figure 3-3   Circuits for the P-N junction and corresponding relation of voltage and current

and negative currents increase with the increasing of the positive and negative voltages. It is found that, when the positive voltage is small, the positive current is also very small. However, when the positive voltage becomes larger than a threshold value, the positive current increases rapidly, indicating that the resistance of the P-N junction is very small with a large positive voltage, as shown by the curve *ob* in Figure 3-3c. In practical situations, if the positive current is too large, the P-N junction will be destroyed. Therefore, we usually set the positive current lower than 25 mA during the experiment. For the case with negative voltage, the equivalent resistance of the P-N junction is very large, and there still exits a threshold value that limits the actual applied voltage on the P-N junction, as shown by the curve *oce* in Figure 3-3c.

The P-N junction could be regarded as a resistance in a circuit that is constructed to measure the voltage and current relation. As mentioned above, the equivalent resistance of the P-N junction is very small with the positive voltage, therefore, we can adopt the ammeter external connection method (shown in Figure 3-2b) to investigate the relation of the current and voltage. However, with the negative voltage, the ammeter internal connection method is introduced (shown in Figure 3-2a) owing to the large equivalent resistance of P-N junction.

IV. Contents and steps

1. Volt ampere characteristic curve

(1) Preparation before constructing the circuit.

i. Adjust the output voltage of the regulated power supply, and make it becomes zero.

ii. Distinguish the high resistance R_{xh} and low resistance R_{xl}.

iii. Understand how to change the measurement range of multi-range ammeter, and know the current and voltage values represented by each small grid for different ranges and the data reading method.

iv. Record the accuracy level of the meters and the inner resistance R_V of the voltmeter with the range from $0\sim3$ V. Record R_A given by the laboratory.

(2) Measure the low resistance R_{xl}.

i. Insert the range plug of voltmeter into the hole that represents the range $0\sim3$ V, and insert the range plug of ammeter in the hole that represents the range $0\sim75$ mA.

ii. The circuit is constructed based on Figure 3-4. Firstly, we place the sliding head of the voltage divider at the minimum voltage output end, then turn on the regulated power supply and set its output voltage as 3 V. Secondly, we turn on the switch K, and adjust the sliding head of the voltage divider until the pointer of the voltmeter points 2.5 V. At the same time, the pointer of the ammeter deflects more than 2/3 of the full scale, otherwise, we should appropriately increase the voltage of the voltmeter. Finally, record the readings of the voltmeter and ammeter.

iii. Set the output voltage of the regulated power supply to be zero. Then, we calculate the low resistance R_{xl} according to formula (3-8) and formula (3-10), and give the expression of the experimental result.

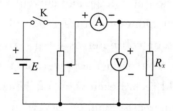

Figure 3-4　Circuit for measurement of low resistance

(3) Measure the high resistance R_{xh}.

i. We change the range of the ammeter to that of $0\sim7.5$ mA, and the range of the voltmeter is set as $0\sim15$ V.

ii. We construct the circuit based on Figure 3-5. The requirements are the same as above. firstly, we adjust the output voltage of the regulated power supply to 15 V, close the switch K, and adjust the sliding head of the voltage divider to make the voltmeter pointer reach about the 14.50 V. In this case, the ammeter pointer should also deflects more than 2/3 of the full scale. Next, we record the readings of the voltmeter and ammeter, and adjust the output voltage of the power supply to zero. Finally, we calculate the high resistance R_{xh} and record the experimental result according to formula (3-8) and formula (3-10).

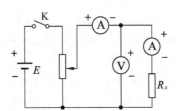

Figure 3-5   Circuit for measurement of high resistance

2. Measurement of volt ampere characteristic curve of crystal diode

The positive current of the 2AP-typed diode should be less than 25 mA, and its negative voltage should be smaller than 14 V.

(1) Measuring the relation of voltage and current with positive voltage.

i. The range of voltmeter is selected as $0\sim3$ V, and that of ammeter is $0\sim30$ mA.

ii. Firstly, we construct the circuit based on Figure 3-6a and adjust the output voltage of the power supply to 2 V. Secondly, we need to check the circuit and turn off the switch K. Finally, we should measure the current by increasing the voltage from 0 V with a step of 0.05 V, until the current reaches 25 mA.

(2) Measuring the relation of voltage and current with negative voltage.

Firstly, we use a micro-ammeter with a range of $0\sim100$ μA to instead the ammeter, and set the range of voltmeter as $0\sim15$ V, as shown in Figure 3-6b. Secondly, we set the output voltage of the power supply as 15 V, and turn off the switch K. Finally, we should record each value of the current corresponding to the voltage of the P-N junction increases from $0\sim2$ V with a step of 0.5 V, then record with a step of 2 V, until the voltage reaches 14 V.

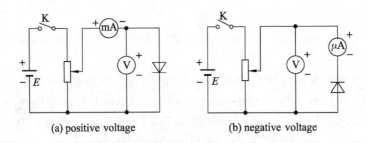

(a) positive voltage   (b) negative voltage

Figure 3-6   Circuits for measurement of volt-ampere characteristic curve of crystal diode with positive and negative voltages

V. Data record and process

1. Measure the low resistance R_{x1}

Voltmeter range $V_m =$ _____ V, inner resistance of voltmeter $R_V =$ _____ Ω,

ammeter range $I_m =$ _____ mA, voltmeter accuracy class $K =$ _____, ammeter accuracy class $K =$ _____ ,

measurement values of the current and voltage: $I =$ _____ mA, $U =$ _____ V,

According to formula (3-8) and (3-10), calculate R_{X1}.

2. Measure the high resistance R_{Xh}

Inner resistance of ammeter $R_A =$ _____ Ω, ammeter accuracy class $K =$

_____ ,

accuracy level of voltmeter $K =$ _____ ,

measured values of voltmeter and ammeter: $U =$ _____ V, $I =$ _____ mA.

The data process is the same as the above step.

3. Measurement of volt ampere characteristic curve of crystal diode

(1) Data for measurement of the voltage and current relation with positive voltage.

The measured values are recorded in Table 3-4.

Table 3-4   Measured voltages and currents for positive voltage

U/V										
I/mA										

(2) Data for measurement of the voltage and current relation with negative voltage.

The measured values are recorded in Table 3-5.

Table 3-5   Measured voltages and currents for negative voltage

U/V										
$I/\mu A$										

VI. Questions

(1) How to measure linear resistance by graphic method?

(2) If give you a SPDT (Single Pole Double Throw) switch, how to combine the two circuits for measuring high and low resistance? Try to draw the circuit diagram and explain the measurement method.

(3) If the resistance is measured by the least square method, what data should be measured? Based on these data, how to calculate the resistance?

Experiment 4　Measurement of the focal length

of thin convex lens

I. Purpose

(1) To study and master the adjustment method of coaxial optical elements.

(2) To master the method of measuring the focal length of thin lens.

(3) To observe the imaging law of convex lens and know its imaging characteristics.

II. Instruments

Optical bench, light source, convex lens, concave lens, small plane mirror, white screen, slide, lamp.

III. Principle

1. Calculate the focal length of the lens according to the object distance and image distance

Set the second main focal length of the thin convex lens as f', an object is placed in front of the lens, and the object distance is s. An image of the object is formed through the convex lens, and the image distance is s'. The relationship between the focal length, object distance and image distance is

$$\frac{1}{s'} - \frac{1}{s} = \frac{1}{f'} \tag{3-11}$$

Therefore, the focal length is calculated by

$$f' = \frac{ss'}{s - s'} \tag{3-12}$$

For formula (3-12), we should define uniform symbol rules which are listed as follows:

i. The intersection point of the light and main axis is measured from the center of the thin lens.

ii. Along the direction of the incident light, the distance between them is positive. Against the direction of the incident light, the distance between them is negative.

iii. Symbols must be added to known quantities during the calculation, while those of unknown quantities are judged by symbols of the calculated results.

According to the geometric optics, when the object is placed within the front focus of the convex lens, the image distance of the object is difficult to be measured. Therefore, the object must be placed outside the front focus of the convex lens in the experiment. In this way, an inverted image will form through the lens. If we put a white screen at position of the image, we can observe it clearly, and the object distance and image distance can be measured directly.

2. Measure the focal length of the lens by conjugation method

We place a convex lens between the object and screen, and keep the distance between the object and screen constant that greater than 4 times of the focal length. By changing the position of the lens between the object and screen, we can find two positions at which the object can be imaged on the screen. As shown in Figure 3-7, PQ is an object, and the distance from PQ to screen is l ($l > 4f'$). When the lens is located at O and O', the images on the screen labeled as $Q'P'$ and $Q''P''$ are observed, respectively. If the distance between OO' is d, the focal length of the convex lens is expressed as

$$f' = \frac{l^2 - d^2}{4l} \tag{3-13}$$

Therefore, we can calculate the focal length by the measured l and d.

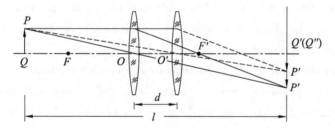

Figure 3-7　Schematic of light path for the conjugation method

3. Measure the focal length of the lens by autocollimation method

As shown in Figure 3-8, the object PQ locates at the left side of the convex lens, and a mirror M is placed at the right side. The light beams emitted from different point on PQ are transformed into parallel beams in different directions after refracted by the convex lens. The light beams will be reflected by the mirror M and transmit through the convex lens again, and form an inverted real image P' Q' at the left side. We can adjust the distances between PQ, convex lens and the mirror M, until the image $P'Q'$ locates at the same position and has the same size of PQ. In this case, the distance between PQ and lens just equals the focal length of the convex lens.

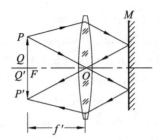

Figure 3-8　Diagram of optical path for autocollimation method

IV. Contents and steps

1. Measure the focal length of lens by object distance and image distance method

（1）Place the object, lens and screen on the light bench in left to right order. The object is a slide illuminated by the light source. Then we adjust the positions of the centers of object, lens and screen carefully, keep them on the same axis parallel to the horizontal direction.

（2）Keep the object at a proper distance from the lens, then adjust the position of the screen until the image on the screen could be observed most clearly. The position of the object, lens and screen could be read from the ruler on the light bench. Finally, we can calculate the object distance and image distance, and substitute them into equation （3-12） to calculate the focal length of the lens.

（3）Change the distance between the object and the lens, repeat step （2） twice, and find the average value of the focal length.

2. Measure the focal length of the lens by conjugation method

（1）Firstly, we arrange the object, lens and screen on the light bench according to Figure 3-7, then adjust the positions of the centers of object, lens and screen carefully, make them satisfy the coaxial condition. The distance between the object and screen is set greater than 4 times of the focal length, fix the object and screen.

（2）Make the image on the screen become clear and enlarged by adjusting the position of the lens, and read the position O of the lens from the ruler on the light bench. Then we move the lens again to get a clear and reduced image on the screen, and read the lens position O'.

（3）Read the position Q of the object and the position Q' of the screen, calculate d and l, then substitute them into equation （3-13）, calculate the focal length.

（4）Change the distance l, repeat the steps （2） and （3） three times, and calculate the average value of the measured focal length.

V. Data record and process

1. Measure the focal length by object distance and image distance method

The measured positions are recorded in Table 3-6.

Table 3-6   Data record for object and image distance method

Times	Position x/mm			s	s'	f'	$\overline{f'}$
	Q	O	Q'				
1							
2							
3							

2. Measure the focal length by conjugation method

The measured values are recorded in Table 3-7.

Table 3-7   Data record for conjugation method

Times	Position x/mm		l	Position x/mm		d	f'	$\overline{f'}$
	Q	Q'		O	O'			
1								
2								
3								

3. Measure the focal length by autocollimation method

The measured values are filled in Table 3-8.

Table 3-8   Data record for autocollimation method

Times	Position x/mm			s	s'	f'	$\overline{f'}$
	Q	O	Q'				
1							
2							
3							

VI. Questions

(1) How to measure the focal length of lens by "autocollimation method"? Draw the schematic of the optical path.

(2) If there is a convex lens whose focal length is greater than the length of the light bench, try to design a scheme that can measure its focal length on the light bench and draw the light path diagram.

Chapter 4

Basic experiment

 Experiment 1   Experiments on air cushion track

I. Purpose

(1) To learn the usage of the digital timer counter anemometer.

(2) To observe the uniform linear motion, and measure the speed and acceleration.

(3) To verify the law of conservation of momentum in both perfect elastic collision and completely inelastic collision.

II. Instruments

Air cushion track (including two optical-electrical doors, two sliders, one cushion, a hasp mode of nylon), digital timer counter anemometer, vernier caliper, several weights.

III. Principle

1. Principle of air cushion guideway

The air cushion track is composed of one steel track and one air pump. There are many small hollows on the surface of steel track. When the pump works, the air will be compressed heavily and flow into the steel track. The air ejected from the small hole on the surface of the steel track forms a very thin air layer (called "air cushion") between the sliders and the steel track. In this case, the contact friction between the sliders and the steel track is avoided, and there exists only a tiny viscous resistance arising from the frictions between the air layers on steel track and the surrounding environment.

Many mechanical laws can be verified on the air cushion along the steel track, such as Newton's first law of motion, Newton's second law of motion, law of

42

conservation of momentum, etc.

2. Measurement of speed

If we fix a light blocking sheet with one leaf on the top of the slider, the light blocking sheet will cut off the optical-electrical door if the slider passes by it. Then, the digital timer counter anemometer will start timing. When the slider passes by the second optical-electrical door, the light blocking sheet will cut off the light path again, and the digital timer counter anemometer will stop timing. In this way, we can measure the time of the slider moving a distance S between the two optical-electrical doors. Therefore, we can calculate the average speed of the slider by

$$\bar{v} = \frac{S}{t} \tag{4-1}$$

If the slider is equipped the light blocking sheet with double leaves, after the slider passing through the optical-electrical door, the digital timer counter anemometer will display the time Δt experienced by the slider movement, and the average speed is

$$\bar{v_t} = \frac{\Delta S}{\Delta t} \tag{4-2}$$

However, the instantaneous speed is given by

$$v_t = \lim_{\Delta t \to 0} \frac{\Delta S}{\Delta t} \tag{4-3}$$

Here, the distance between the two leaves is ΔS ($\Delta S \approx 10.00$ mm), which could be considered as a very small displacement comparing with the length of the steel track. Therefore, we finally have

$$v_t \approx \bar{v_t} = \frac{\Delta S}{\Delta t} \tag{4-4}$$

3. Measurement of uniform acceleration

By adding a cushion under the left feet of steel track, the air cushion will not be horizontal. In this case, the steel track becomes an inclined plane with a negligible friction. If we place a slider at the left end of the air cushion, it will move from the left to right with an uniform acceleration which induced by the component of the gravity along the track. Now, two optical-electrical doors are placed at different positions on the track, the distance between them is S. The instantaneous speeds of the slider at the two optical-electrical doors could be measured according to formula (4-4), which are recorded as v_1 and v_2, respectively. The acceleration of the slider along the track is given by

$$a = \frac{v_2^2 - v_1^2}{2S} \tag{4-5}$$

4. Verify the law of conservation of momentum

For a system without any external force or combined force, the total momentum of the system remains unchanged, which is called the law of conservation of momentum. Suppose that the system contains only two objects and they collide with each other along a straight line, the momentum of the system in the collision direction will be conserved only when the total components of the combined forces in the same direction is zero.

In this experiment, the masses of two sliders are m_1 and m_2, respectively. Before their collision along the track, their speeds are v_{10} and v_{20}. After the collision, their speeds become v_1 and v_2 respectively. According to the law of conservation of momentum, we have

$$m_1 v_{10} + m_2 v_{20} = m_1 v_1 + m_2 v_2 \tag{4-6}$$

(1) Perfect elastic collision.

In this situation, the kinetic energy is also conservation. So, we have

$$\frac{1}{2} m_1 v_{10}^2 + \frac{1}{2} m_2 v_{20}^2 = \frac{1}{2} m_1 v_1^2 + \frac{1}{2} m_2 v_2^2 \tag{4-7}$$

Using formula (4-6) and (4-7), we will obtain

$$v_1 = \frac{(m_1 - m_2) v_{10} + 2 m_2 v_{20}}{m_1 + m_2}$$

$$v_2 = \frac{(m_2 - m_1) v_{20} + 2 m_1 v_{10}}{m_1 + m_2} \tag{4-8}$$

Especially, when $v_{20} = 0$, and $m_1 = m_2 = m$, formula (4-8) becomes $v_1 = 0$, $v_2 = v_{10}$.

(2) Completely inelastic collision.

In this situation, only the momentum of the system is conserved.

For $v_1 = v_2 = v$, according to formula (4-6), we can obtain that

$$v = \frac{m_1 v_{10} + m_2 v_{20}}{m_1 + m_2} \tag{4-9}$$

When $v_{20} = 0$, and $m_1 = m_2 = m$, we will get $v = \frac{1}{2} v_{10}$.

IV. Contents and steps

1. Preparation

First, open the pump, and put the slider on the steel track. Then we adjust the screws at the bottom of steel track, until the slider keeps static on steel track. In this situation, the steel track is horizontal.

2. Measurement of the average speed

(1) Fix two optical-electrical doors at different positions on the steel track, and the distance S between them is set as 600.0 mm.

(2) Open the digital timer counter anemometer, and select the "S_2" function.

(3) Measure the thickness of the cushion by using vernier caliper, and put it under the left foot of the steel track.

(4) Add the light blocking sheet with one leaf on the top of the slider, place the slider on the left end of steel track and keep it static, and then release it.

(5) The slider will slide along the track under the action of gravity, and pass by the two optical-electrical doors successively. Record the time that the slider spends moving from the first to second doors.

(6) Repeat steps (4) and (5) for six times. Note that, the slider should be put on the same position on the steel track for each time.

3. Measurement of acceleration

(1) Fix the light blocking sheet with two leaves on the top of the slider.

(2) Put the slider on the left end of the steel track, and make it slide along the track freely.

(3) When the slider passes through first and second optical-electrical doors successively, read two data Δt_1 and Δt_2 from the displayer of the digital time counter anemometer respectively.

(4) Repeat steps (2) and (3) six times.

4. Verification of the law of conservation of momentum

(1) Remove the cushion, and keep the track horizontal.

(2) Put a slider with mass of m_1 between two optical-electrical doors, and keep it static.

(3) Put the other slider with mass of m_2 at the left end of the steel track, push it and make it move towards the first slider.

(4) After collision of the two sliders, we also can read two data Δt_1 and Δt_2 from the displayer of the digital time counter anemometer.

(5) Note that, one end of the slider is added with a spring, and the other end is attached with nylon.

For the case of perfect elastic collision, we just let the two sliders collide with the spring ends, while for the case of completely inelastic collision, we should let the two sliders collide with the nylon ends.

(6) Repeat the measurements for three times.

V. Data record and process

1. Measurement of the average speed

$S =$ _____ mm

The measured values are filled in Table 4-1.

Table 4-1 Data record for the measurement of average speed

Times	1	2	3	4	5	6	$\bar{t}$
t_i/s							

$$\bar{v} = \frac{S}{\bar{t}} = \text{_____}$$

2. Measurement of acceleration

Interval distance between two leaves light blocking sheets $\Delta S =$ _____ mm, thickness of cushion $d =$ _____ mm, effective length of inclined plane $L = 860.0$ mm.

The measured t_1 and t_2 are recorded in Table 4-2.

Table 4-2 Data record for the measurement of acceleration

Times	t_1/s	t_2/s
1		
2		
3		
4		
5		
6		
Average		

$$v_1 = \frac{\Delta S}{\bar{t}_1} = \text{_____}, \quad v_2 = \frac{\Delta S}{\bar{t}_2} = \text{_____}, \quad a = \frac{v_2^2 - v_1^2}{2S} = \text{_____}.$$

The theory value $a = g \cdot \sin \alpha = g \cdot \dfrac{d}{L} = $ _____.

3. Verification of the law of momentum conservation

$m_1 =$ _____ g, $m_2 =$ _____ g, $\Delta S_1 =$ _____ mm, $\Delta S_2 =$ _____ mm.

(1) Perfect elastic collision.

The measured Δt_1 and Δt_2 are filled in Table 4-3.

Table 4-3 Data record for perfect elastic collision

Times	$\Delta t_1/s$	$v_{10}/$ (mm $\cdot$ s^{-1})	$\Delta t_2/s$	$v_2/$ (mm $\cdot$ s^{-1})	$m_1 v_{10}/$ (g $\cdot$ mm $\cdot$ s^{-1})	$m_2 v_2/$ (g $\cdot$ mm $\cdot$ s^{-1})
1						
2						
3						

(2) Completely inelastic collision.

The measured values are filled in Table 4-4.

Table 4-4   Data record for completely inelastic collision

Times	$\Delta t_1/s$	$v_{10}/$ $(mm \cdot s^{-1})$	$\Delta t_2/s$	$v_2/$ $(mm \cdot s^{-1})$	$m_1 v_{10}/$ $(g \cdot mm \cdot s^{-1})$	$(m_1+m_2)v_2/$ $(g \cdot mm \cdot s^{-1})$
1						
2						
3						

VI. Questions

(1) How to adjust and judge whether is the steel track horizontal?

(2) When we measure the acceleration, why ΔS is not the width of the light blocking sheet or the slit width of the double light blocking sheet?

(3) When comparing the acceleration measured in the experiment with the theoretical value $\left(a=g \cdot \sin \alpha=g \cdot \dfrac{d}{L}\right)$, it is generally found that the experimental value is small. What does this kind of error belong to? What are the causes? Which methods can be used to eliminate or reduce the error?

(4) In the case of elastic collision, when $m_1=m_2=m$ and $v_{20}=0$, please use the measured data to calculate and verify whether the total energy before and after collision is equal. If not, please analyze the reasons.

Experiment 2 Measurement of Young's modulus

I. Purpose

(1) To observe the law of the elastic deformation of the steel wire under the stress along the axial direction.

(2) To measure the Young's modulus of the steel wire with the static tensile method.

(3) To understand the principle of the optical lever, and know how to measure the small variation of the steel-wire's length.

(4) To study the successive difference method.

II. Instruments

Young's modulus tensile apparatus, optical lever, weights, steel tape, screw micrometer.

III. Principle

Any material in nature can be elastically stretched or compressed, in which the deformation variation is proportional to the intensity of the external tensile force. The concept of Young's modulus, originally proposed by a famous British physicist Thomas Young, can well describe the elastic properties of the materials. In this experiment, the Young's modulus of the steel wire is obtained by measuring the tiny elongation of the steel wire with an axial tensile force.

Figure 4-1 shows the deformation of a vertically suspended steel wire, in which L is the initial length, S is the cross sectional area, and ΔL is the elongation of the steel wire induced by the axial tensile force F. The Young's modulus E along the axial direction is defined as

$$E = \frac{\sigma}{\varepsilon} \tag{4-10}$$

where the stress σ and strain ε are given by $\sigma = F/S$ and $\varepsilon = \Delta L/L$, respectively. It has been proved that Young's modulus was only associated with the essential properties of the materials and independent of the geometric shape of the materials. From equation (4-10) and Figure 4-1, the Young's modulus can be calculated by

$$E = \frac{F \cdot L}{S \cdot \Delta L} \tag{4-11}$$

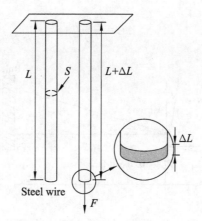

Figure 4-1   Schematic of the elongation of the steel wire with an axial tensile force F

However, the elongation ΔL of the steel wire is too tiny to be measured directly. Commonly, it is very efficient to enlarge the tiny mechanical deformations (stretching or compressing variations and tiny rotation angle) by using optical lever in physics experiment.

Figure 4-2 shows the enlarging principle of the optical lever. The system consisted of a telescope and a ruler is located at a distance D' away from the mirror. The telescope faces to the mirror so as to observe the image of the ruler in the mirror conveniently. Before stretching the steel wire, the mirror stands on the platform vertically. In this situation, the light of the sight will be reflected back directly by the mirror along the incident direction, and then travel into the eyepiece of the telescope. Therefore, the image of the ruler is observed in the eyepiece (seeing the insert on the lower right panel in Figure 4-2), and the reading data is recorded as Y_0. Once the steel wire is stretched with an elongation ΔL, the measuring end-face as well as the rear fulcrum of the mirror will sink down with the same distance of ΔL. As a result, the mirror rotates a tiny angle of θ around the axis perpendicular to the panel of Figure 4-2. From the principle of reflection, the light of sight reflected by the rotated mirror will propagate along the direction deviated away from the incident direction with an angle of 2θ. Thus, the reading data through the eyepiece will be changed to Y_i. Derivable from the trigonometric function relationship in Figure 4-2, we have

$$\theta \approx \frac{\Delta L}{l} \tag{4-12}$$

$$2\theta \approx \frac{b}{D'} \tag{4-13}$$

where l is the length of the rear fulcrum of the mirror, and θ is very small. Based on equations (4-12) and (4-13), we obtain

$$\Delta L = \frac{l}{2D'}b \qquad (4\text{-}14)$$

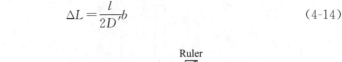
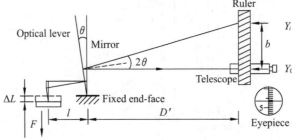

Figure 4-2　Schematic of the optical enlarging of the tiny elongation ΔL by the optical lever

To increase the enlarging times of the optical lever, we can move the ruler to the side of the telescope on the fixed endface, and place another reflected mirror beside the telescope, as shown in Figure 4-3. The light of sight will be reflected twice between two mirrors, and finally travel along the direction deviating an angle of 4θ from the incident direction. According to Figure 4-3, we have

$$\Delta L = \frac{l}{4D'}b \qquad (4\text{-}15)$$

By substituting equation (4-15) into equation (4-11), the Young's modulus is calculated by

$$E = \frac{16FLD'}{\pi d^2 lb} \qquad (4\text{-}16)$$

in which the cross-sectional area of the steel wire is given by $S = \pi d^2/4$, and d is diameter of the steel wire.

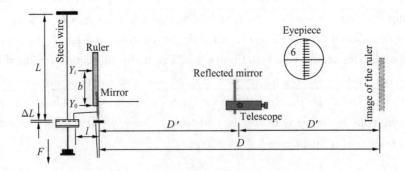

Figure 4-3　Schematic of the measurement of ΔL

IV. Contents and steps

(1) Make the center axis of the telescope be horizontal which is also perpendicular to the surface of two mirrors.

(2) Adjust the vertical position of the telescope until it is almost the same as that of the optical lever.

(3) Adjust the eyepiece of the telescope, until you can see the graticule in the eyepiece clearly, as shown the inserts on the left panel of Figure 4-4.

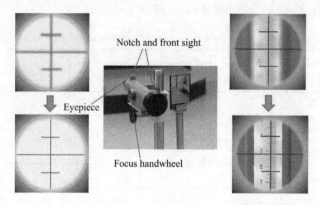

Figure 4-4   Schematic of the adjustment of the telescope

(4) Make the notch and the front sights on the telescope aim at the image of the ruler in the mirror of the optical lever, and then try to find the image of the ruler in the eyepiece of the telescope (seeing the insert on upper panel of Figure 4-4). Adjust the focus handwheel at the side of the telescope to make the image of the ruler in eyepiece clear (seeing the insert on lower right panel of Figure 4-4). It is worth to note that we can record the position data from the image of the ruler according to the middle horizontal graticule.

(5) Put a weight with 100 g on the pallet. According to the lever principle, the load at the lower end of the steel wire is 1 kg. Then record the reading data as Y_1 from the image of the ruler in the eyepiece. Add the weight with 100 g in turn, until the number of the weights reaches eight. The loads of the steel wire are 2 kg, 3 kg, 4 kg, 5 kg, 6 kg, 7 kg and 8 kg, and we should read the data from the image of the ruler in the eyepiece and record them as Y_2 to Y_8, respectively.

(6) Remove the weights on the pallet in turn, and record the reading data, respectively. Note that for both the cases of increasing and decreasing the weights, the measured data should nearly equal with each other if they correspond to the same load of the steel wire. Otherwise, we should check the operation in the experiment and try to find the causes of the problem.

(7) Measure the distance D between the ruler and the reflection mirror and the initial length of the steel wire by using the steel tape.

(8) Measure the diameter of the steel wire by using screw micrometer. The measurements are performed twice at the upper, middle and lower part of the steel

wire, respectively. Then we choose the mean value of the six results as the measured diameter d.

V. Data record and process

1. Data for the measurement of $\bar{b}$

The measured values are filled in Table 4-5.

Table 4-5   Data record for the measurement of $\bar{b}$

Times	Load/ kg	Adding Y_i/mm	Removing Y_i/mm	Mean value Y_i/mm	$b_i = Y_{i+4} - Y_i$/mm
1	1.00				
2	2.00				
3	3.00				
4	4.00				
5	5.00				
6	6.00				$\bar{b} =$
7	7.00				
8	8.00				

$L =$ _____ mm, $D =$ _____ mm.

2. Data for the measurement of d

The measured values are recorded in Table 4-6.

Table 4-6   Data record for the measurement of the diameter of steel wire

Times	1	2	3	4	5	6	Average
d_i/mm							

$$\bar{E} = \frac{16D'FL}{\pi d^2 l \bar{b}} = \underline{\qquad} \; N/m^2. \; (F = 4 \times 9.8 \; N)$$

VI. Questions

(1) In the experiment, we calculate the Young's modulus by equation (4-11). What conditions are needed to make the equation (4-11) valid?

(2) Please calculate the Young's modulus E by drawing figure in computer.

(3) How to improve the measurement accuracy of ΔL by optical lever?

Experiment 3   Measurement of the moment of inertia
by three-wire pendulum

I. Purpose

(1) To measure the moment of inertia by three-wire pendulum.

(2) To verify the parallel axis theorem.

II. Instruments

FB210 three-wire pendulum, FB213 digital millisecond counter, vernier caliper, meter ruler, steel ring, two steel cylinders.

III. Principle

In classical mechanics, the moment of inertia is a measure of the resistance to change the rotational motion of an object, describing the relationship between angular momentum, angular velocity, moment and angular acceleration.

For a particle with mass of m_i, the moment of inertial is calculated as $I_i = m_i r_i^2$, where r_i is the vertical distance between the particle and the axis of rotation. The moment of inertia of an object considered as the collective of particles is given by

$$I = \sum_i I_i = \sum_i m_i r_i^2 \tag{4-17}$$

If the object is continuous, the summation in equation (4-17) will become integration.

However, in many practical situations, the complex and irregular shape of the object will make it very hard or impossible to calculate the moment of inertia directly. Therefore, it is very essential to introduce an approach to measure the moment of inertia. In the experiment, we will show the measurement of the moment of inertial by three-wire pendulum.

The three-wire pendulum is composed of a small horizontal disc fixed on the top and a suspended uniform disc by three wires, as shown in Figure 4-5. The rotation motion of the suspended disc around the central axis with a small angle amplitude could be considered as simple harmonic vibration. Ignore the resistance, from the law of conservation of mechanical energy and simple harmonic vibration, the moment of inertia of the suspended disc around central axis is expressed as

$$J_0 = \frac{m_0 g R r}{4\pi^2 H} T_0^2 \tag{4-18}$$

where m_0 is the mass of the suspended disc, g is the gravitational acceleration, r

and R are the radius of the small horizontal disc and suspended disc respectively, H is the distance between the two discs, and T_0 is the period of the rotation motion of the suspended disc.

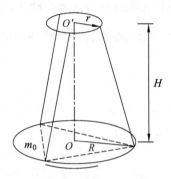

Figure 4-5   Schematic of the three-wire pendulum

When a steel ring with the mass m_1 is placed on the suspended disc and keep the central axes of the ring and disc coincide with each other, the moment of inertia of the system becomes

$$J = \frac{(m_0 + m_1)gRr}{4\pi^2 H}T_1^2 \tag{4-19}$$

where T_1 is the period of the rotation motion of the system. The moment of inertia of the steel ring around the central axis is

$$J_1 = J - J_0 \tag{4-20}$$

Similarly, when two identical steel cylinders are placed on the suspended disc symmetrically about the central axis, the moment of inertia of the two cylinders is

$$J_2 = \frac{(m_0 + 2m_2)gRr}{4\pi^2 H}T_2^2 - J_0 \tag{4-21}$$

here, m_2 is the mass of the steel cylinder, T_2 is the period of the rotation motion of the suspended disc with two steel cylinders.

IV. Contents and steps

(1) Adjust the screws at the bottom of the instrument to make the small disc at the top be horizontal, then adjust the screws at the top of the pendulum until the bubble moves into the center circle of the bubble level and becomes static. Then tighten the screws at the top end of the suspended lines.

(2) Adjust the position of the photoelectric gate to avoid the collision of the photoelectric gate and shading rod, which will rotate with the rotation of the suspended disc.

(3) Set the measurement times as 50 by adjusting the preset key of the FB213 digital millisecond counter.

(4) Make the suspended disc rotate around its central axis with a small angle by pushing the handle of the small disc on the top. After several periods of the rotation of the suspended disc, press the starting key to start the measurement. The FB213 digital millisecond counter will finish the measurement automatically once the measurement times reaches the presetting value (50 times). The measured value is recorded as t, and one period of the rotation of the pendulum is $T_0 = t/50$. To decrease the measurement error, we should redo the measurement of T_0 three times, and take the average value as the final result. Note that before redoing the measurement, we should press the reset key to clear the data of the former step.

(5) Locate the steel ring on the suspended disc, and keep the central axes of the ring and disc coincide with each other. Repeat step 4, and record the period of the system as T_1.

(6) Locate the two cylinders on the suspended disc symmetrically about the central axis. Repeat step 4, and obtain the period of the system.

V. Data record and process

Distance between two discs $H = $ _____ mm,

ring' inner diameter $D_{in} = $ _____ mm, outer diameter $D_{out} = $ _____ mm,

bottom disc's diameter $D_1 = $ _____ mm, cylinder's diameter $D_2 = $ _____ mm.

Distances between suspension points on bottom disc:

$a_1 = $ _____ mm, $a_2 = $ _____ mm, $a_3 = $ _____ mm.

Distances between suspension points on top disc:

$b_1 = $ _____ mm, $b_2 = $ _____ mm, $b_3 = $ _____ mm,

$\bar{a} = (a_1 + a_2 + a_3)/3 = $ _____ mm, $\bar{b} = (b_1 + b_2 + b_3)/3 = $ _____ mm,

$R = \bar{a}/\sqrt{3} = $ _____ mm, $r = \bar{b}/\sqrt{3} = $ _____ mm,

mass of bottom disc $m_0 = $ _____ g, mass of ring $m_1 = $ _____ g,

masses of two cylinders $m_{21} = $ _____ g, $m_{22} = $ _____ g.

The measured t are recorded in Table 4-7.

Table 4-7   Data record for the measurement of period T

Times of measurement	1	2	3	Mean value ($\bar{t}$)	$\bar{T} = t/50$
50 times periods t/s					
50 times periods with the ring t/s					

$$J_0 = \frac{m_0 gRr}{4\pi^2 H} T_0^2 = \underline{\hspace{2cm}},$$

$$J_1 = \frac{(m_0 + m_1)gRr}{4\pi^2 H} T_1^2 - J_0 = \underline{\hspace{2cm}}.$$

The measured distance d and time t are filled in Table 4-8.

Table 4-8　Data record for the measurement of distance between two cylinders and the time t

Distance between cylinders $2d$/mm					
d^2/mm^2					
50 times periods $\bar{t}$/s					
Period T/s					
Inertia moment J_2					

Drawing the figure, and verify the parallel axis theorem.

VI. Questions

(1) Why is it necessary to keep the lower disk horizontal when measuring the moment of inertia of a rigid body with a three-wire pendulum?

(2) During the measurement process, the lower disk shakes, does it influence the measurement of the period? If so, how to avoid it?

(3) Can the parallel axis theorem be verified by graphic method in this experiment? How to verify it?

(4) How to use a three-wire pendulum to measure the moment of inertia of an object with arbitrary shape around a specific axis?

(5) In addition, is the period of the three-wire pendulum with an object necessarily larger than that of the empty disk? Why?

(6) Three-wire pendulum is damped by air resistance, and its amplitude becomes smaller and smaller. Will its period change? Does it have a great impact on the measurement results? Why?

Experiment 4   Measurement of thermal conductivity of poor conductor by steady-state method

I. Purpose

(1) To understand the physical law of the transmission of the heat.

(2) To master the measurement of temperature by the technology of thermocouple.

(3) To learn the measurement of thermal conductivity of undesired conductors by steady-state method.

II. Instruments

A set of thermal conductivity tester (including voltage regulator, heating plate, rubber sample to be tested, sample support, thermocouple, heat dissipation fan, Dewar bottle, digital voltmeter, etc.), vernier caliper, balance.

III. Principle

1. Heat conduction

When the distribution of the temperature inside an object is not uniform, the heat will be automatically transferred from the part with higher temperature to that with lower temperature. This phenomenon is the heat transmission. The physical law of the heat transmission was first proposed by the famous physicist Fourier in 1882, which is described as follows: if a section ΔS is taken perpendicular to the heat transmission direction, the heat $\Delta Q/\Delta t$ (also known as heat conduction rate) transferred from the higher temperature side to the lower temperature side in unit time is proportional to the temperature gradient dT/dx, namely

$$\Delta Q/\Delta t = -\kappa(dT/dx)\Delta S \qquad (4\text{-}22)$$

Formula (4-22) is called Fourier heat transmission equation, where the scaling coefficient κ is called heat conductivity or coefficient of conductivity with the unit $W/(m \cdot K)$. The negative sign in the formula represents the direction of heat transmission opposite to the temperature gradient.

2. Measurement of the heat conductivity by steady-state method

Figure 4-6 shows the schematic of the steady-state insandwich structure, in which B is a rubber plate to be measured, A is the heating plate, and C is the cooling plate made of copper, D is the diameter of the plate A, B and C, h is the thickness of B, T_1 and T_2 are the temperatures at the lower surface of A and the upper surface of C, respectively. The heat from the heating plate A is transferred to the rubber plate B, and then flows into the cooling plate C. In the course of time, the system reaches a steady-state, in which the temperatures on the upper and lower surfaces T_1 and T_2 do not change with the time. Inside of the rubber plate B, the temperature gradient is established and remains stable. Note that, as shown in Figure 4-6, the diameter of the plate A, B and C is much larger than the thickness of the rubber plate B ($D \gg h$), which make the heat dissipation from the side of the plate B be ignored. Therefore, formula (4-22) can be written as

$$\Delta Q/\Delta t = -\kappa[(T_1 - T_2)/h] \cdot \Delta S \qquad (4\text{-}23)$$

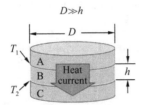

Figure 4-6　Schematic of the steady-state insandwich structure

When the heat transmission in plate B reaches the steady state, the heat transmission rate from the lower surface of the plate B to the cooling plate C equals to the heat dissipation rate of the copper plate C. And the heat dissipation rate can be measured by the temperature cooling rate of the copper plate under the same external conditions at the temperature T_2.

According to the heat equation $\Delta Q = cm\Delta T$, we can obtain the temperature cooling rate of the copper plate at temperature T_2, namely

$$(\Delta Q/\Delta t)|_{T=T_2} = cm(\Delta T/\Delta t)|_{T=T_2} \qquad (4\text{-}24)$$

where c is the specific heat capacity of copper, m is the mass of copper plate C, $\Delta T/\Delta t$ and $\Delta Q/\Delta t$ are the temperature and heat variation rate in copper plate C, respectively.

In the experiment, we can measure the temperature T_1 and T_2 in the steady state, then take away the sample B and make plate A heats plate C directly until the temperature becomes 10 ℃ higher than T_2. Next, remove the heating plate A, cover the cooling plate C with the rubber plate B, and start the natural cooling process of the plate C. Record the temperature of the copper plate C every 30 seconds, and plot the corresponding cooling curve, based on which we can obtain the temperature variation rate of the copper plate C at T_2. Therefore, according to equations (4-22)~(4-24), we obtain that

$$\kappa = 4mch(\Delta T/\Delta t)|_{T=T_2}/[\pi D^2(T_1 - T_2)] \qquad (4\text{-}25)$$

3. Thermocouple principle for temperature measurement

According to the classical electronic theory, different metals have different electronic escape powers. When two metals with different electronic escape powers are attached with each other, a contact potential difference will be created between them. The magnitude of the contact potential difference is related to the temperatures of the two metals. If we place two identical contacting ends of the metals at different temperatures and connect them with wires, a contact potential differences will be created at the interface. We call the total electromotive force in the above closed circuit as thermoelectric force. According to this principle, we can convert the measurement of temperature into that of potential difference. In this experiment, the thermocouples are fabricated by copper-constantan. Commonly, the temperature at one contact end of the thermocouple is fixed as the reference. We select the ice-water mixture as the temperature reference point T_0 (0 ℃).

The thermoelectric force is proportional to the temperature difference, namely $\Delta \varepsilon = \alpha \Delta T$. Substitute it into equation (4-25), we have

$$\kappa = 4mch (\Delta \varepsilon / \Delta t)|_{\varepsilon = \varepsilon_2} / [\pi D^2 (\varepsilon_1 - \varepsilon_2)] \tag{4-26}$$

IV. Contents and steps

(1) Select the 20 mV gear of the digital voltmeter. Firstly, we zero the digital voltmeter by pressing the 20 mV range switch. Secondly, we make the input end short circuit, and press the zero-adjustment switch. Next, adjust the zero-adjustment knob to make the digital voltmeter display "000".

(2) Put the ice water mixture into the Dewar bottle. Do not remove or reconnect the thermocouple by yourself to avoid damaging the device.

(3) Note that, it will cost about 1 hour to make the system reachs the steady state. To shorten the time, we increase the output voltage to 220 V firstly. After a few minutes, when the displayed value of the temperature difference thermocouple on the upper surface of the sample is $\varepsilon_1 = 4.00$ mV, decrease the voltage to 110 V. When ε_1 drops to about 3.50 mV, alternately change the voltage to 220 V, 110 V and 0 V to make the ε_1 within the range of (3.50 ± 0.03) mV. At the same time, check the voltage ε_2 on the lower surface of the sample every 2 minutes. It can be considered that the steady state has been reached when ε_2 remains unchanged within 10 minutes, and record the values of ε_1 and ε_2, respectively.

(4) In this step, we take away the sample and make the heating plate heat the cooling plate directly. We only measure the temperature of the cooling plate. When the temperature of the cooling plate increases about 10 ℃ (the temperature difference electromotive force is about 1 mV higher than ε_2), remove the heating plate, cover the cooling plate with the sample (plate B), let the system cool

naturally, and measure the temperature of the cooling plate every 30 seconds until the value of the digital voltmeter is 0.5 mV lower than ε_2.

(5) Measure the diameter D and thickness h of the sample plate with a vernier caliper, and record the mass of the cooling plate.

(6) Plot the ε-t curve by using the time t as the x-axis and temperature difference electromotive force as the y-axis. The slop of the tangent at ε_2 is the value of $d\varepsilon/dt\,|_{\varepsilon=\varepsilon_2}$.

V. Data record and process

(1) Data record.

Diameter of sample plate $D=$ _____ mm, thickness $h=$ _____ mm,

mass of cooling plate $m=$ _____ g,

specific heat capacity of copper $c=380$ J/(kg $\cdot$ K),

display value of temperature difference thermocouple in steady state:

$\varepsilon_1=$ _____ mV, $\varepsilon_2=$ _____ mV.

The measured values of ε are filled in Table 4-9.

Table 4-9　Data record for the measurement of temperature difference electromotive force

t/s	0	30	60	90	120	150	180	210	...
ε/mV									

(2) Plot the ε-t carve, calculate the value of $d\varepsilon/dt\,|_{\varepsilon=\varepsilon_2}$. Calculate the thermal conductivity based on formula (4-26).

VI. Questions

(1) Try to explain the meaning of the steady state method used in this experiment.

(2) Why should the value be selected near the stable temperature T_2 (electromotive force display value ε_2) when measure the cooling rate?

Experiment 5　Physical properties of P-N junction and measurement of Boltzmann constant

I. Purpose

(1) To be familiar with the physical characteristics of P-N junction.

(2) To master the method of finding empirical formula through data process.

II. Instruments

P-N junction measurement instrument, TIP31 triode (with three leads), glass tube containing a small amount of transformer oil, thermos cup, thermometer.

III. Principle

According to semiconductor physics, the relationship between forward current and voltage of P-N junction is given by

$$I = I_0 \left[e^{e_0 U/(kT)} - 1 \right] \tag{4-27}$$

where I is the forward current through the P-N junction, I_0 is a constant that does not vary with the voltage, k is the Boltzmann constant, T is the temperature, e_0 is the basic charge of one electron, and U is the forward voltage of P-N junction.

Since $kT/e_0 \approx 0.026$ V at room temperature (about 300 K) and the forward voltage of P-N junction is only about a few tenths of volts, $e^{e_0 U/(kT)} \gg 1$, therefore, equation (4-27) can be rewritten as

$$I = I_0 e^{e_0 U/(kT)} \tag{4-28}$$

The formula indicates that the forward current of P-N junction varies exponentially with the forward voltage. If the I-U relationship of P-N junction is measured, $e_0/(kT)$ can be obtained by using equation (4-28). If we know the temperature T, the constant e_0/k can be obtained. The Boltzmann constant k can be obtained by substituting the basic charge e_0 as a known value.

Actually, although the positive I-U relationship of the diode can well meet the exponential relationship, the constant k obtained is often smaller than the standard one. Because the current passing through the diode is not only diffusion current, but also includes other currents which are listed as follows:

(1) Diffusion current, which strictly follows formula (4-28).

(2) Depletion layer recombination current, which is proportional to $e^{e_0 U/(2kT)}$.

(3) Surface current, which is induced by the impurities at the interface between Si and SiO_2, and its value is directly proportional to $e^{e_0 U/(mkT)}$ $(m > 2)$.

Therefore, to verify formula (4-28) and calculate the accurate constant e_0/k, it is not suitable to adopt the silicon diode. Here, we use the silicon triode to construct the common base line, because the collector is short circuited with the base and the collector current is only diffusion current. The silicon triode TIP31 (NPN tube) shows a good performance in the experiment. The experimental circuit is shown in Figure 4-7.

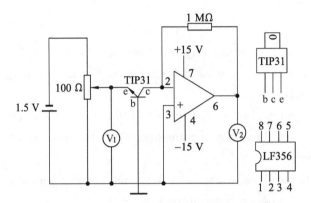

Figure 4-7　Schematic of the experiment circuit

IV. Contents and steps

(1) At room temperature, measure the voltage U_1 between the emitter and the base of the triode and the corresponding output voltage U_2. The value of U_1 increases from 0.300 V with a step of 0.01 V, the corresponding value of U_2 is recorded until the U_2 reaches saturation.

(2) Record the temperature of transformer oil at start and the end of data recording, and take the average temperature T.

(3) Change the water temperature in the thermos cup, stir the water until the temperature in water is consistent with the oil temperature in the glass tube, repeat the measurement in steps 1, and compare it with the results measured at room temperature.

V. Data record and process

1. Measure at room temperature

Oil temperature at start $T_1 = $ _____ K,

oil temperature at the end $T_2 = $ _____ K,

average $T = (T_1 + T_2)/2 = $ _____ K.

The measured U_1 and U_2 for room temperature are filled in Table 4-10.

Table 4-10   Data record for the case of room temperature

U_1/V											
U_2/V											

2. Measure at high temperature

Oil temperature at start $T_1 = $ _____ K,

oil temperature at the end $T_2 = $ _____ K,

average $T = (T_1 + T_2)/2 = $ _____ K.

The measured values for high temperature are filled in Table 4-11.

Table 4-11   Data record for the case of high temperature

U_1/V											
U_2/V											

3. Fit the curve of empirical formula by using the least square method

4. Calculate the Boltzmann constant k

VI. Questions

(1) What problems should be paid attention to when we measure the P-N junction temperature in this experiment?

(2) What are the main errors to be eliminated by connecting the TIP31 triode into a common base circuit and measuring the relationship between diffusion current of the P-N junction and voltage?

Experiment 6　Adjustment and application of oscilloscope

I. Purpose

(1) To understand the basic structure and working principle of oscilloscope.

(2) To master the adjustment of oscilloscope.

(3) To learn the usage of the oscilloscope to analyze the signal.

(4) To master the measurement of signal's frequency by using Lissajous' picture.

II. Instruments

Oscilloscope, functional signal generator, transformer.

III. Principle

1. Structure of oscilloscope

The oscilloscope is consisted with three main parts, namely oscilloscope tube, control circuit and the power. The oscilloscope tube is the core component, it consists of several parts: glass shell, electric gun, deflection system and screen.

(1) The electric gun can emit electron beam with high speed.

(2) Deflection system is composed with two groups of parallel metal plates. One called vertical deflection plates and another is horizontal deflection plates. When we add a voltage on the deflection plates, the electron beam flowing through them will be effected by electric field force, and the propagation direction of the electron beam will be deflected. The deflection value is determined by the magnitude of the voltage added on two metal plates.

(3) The screen is fabricated by coating a layer of fluorescent substance on the inner wall of oscilloscope's bigger end. When the electrons bombard the fluorescent substance, the screen will be luminesced.

The control circuit has several functions. Firstly, it can amplify or attenuate signal's intensity. Secondly, it provides low and high voltages to oscilloscope. Thirdly, it can generate a period saw-tooth wave signal in x-axis to scan the signal in y-axis.

2. Oscilloscope principle

(1) To obtain a real shape of signal in y-axis, we must apply a period saw-tooth wave signal in x-axis. If we set the signal in y-axis as

$$y = A\cos wt \qquad\qquad (4\text{-}29)$$

and its period is T_y, then the signal in x-axis is

$$x=t(0{\leqslant}t{\leqslant}T_x)\qquad\qquad\qquad(4\text{-}30)$$

T_x is the period of signal in x-axis. Only when $T_x=nT_y(n=1,2,3\cdots\cdots)$, the signal in y-axis could be displayed on screen completely and stably.

(2) If we input two sinusoidal signals into the oscilloscope from channels 1 and 2, respectively, we will observe the Lissajous' picture on the screen. If $f_y/f_x=n(n=1,2,3\cdots\cdots)$, the Lissajous' picture is a stable and clear graph, where f_x is the frequency of signal input from channel 1 and f_y is that input from channel 2. In this case, if we draw a vertical line to approach the vertical edge of Lissajous' picture and a horizontal line tangent to the horizontal edge Lissajous' picture, respectively, we can find the numbers of vertical tangent points n_y and horizontal tangent points n_x, respectively. Then we have the relationship between the frequencies and the numbers of the tangent points: $\dfrac{f_y}{f_x}=\dfrac{n_x}{n_y}$. Therefore, if we know f_x, and find the n_x and n_y, we will obtain f_y.

IV. Contents and steps

(1) Open the instrument's power.

(2) Choose channel 1 by press the button CH1.

(3) Observe the sinusoidal signal with frequency 50 Hz which is generated by functional signal generator, and draw the signal on paper.

(4) Observe the sinusoidal signal with frequency 420 Hz and 13 kHz, respectively.

(5) Choose channel 2 by pressing the button CH2.

(6) Adjust the volts/div knob and time/div knob, until a perfect sinusoidal signal appears on the screen. Then record the value of volts/div and time/div, and measure its V_{p-p}'s value.

(7) Press the x-y button to combine signals input from channels 1 and 2 together.

(8) Adjust the frequency of the signal from channel 2 to 25 Hz by rotating the frequency knob on control board of the signal generator, draw the Lissajous' picture on paper. Then adjust the frequency of the signal to 50 Hz, 75 Hz, 100 Hz, and 200 Hz, draw the Lissajous' pictures, respectively.

V. Data record and process

Sine-shaped signal with frequency of 50 Hz,

volts/div = _____, time/div= _____,

alternating current's peak to peak value volts/div $u=$ _____,

number of division line $n=$ _____.

The peak to peak value $V_{p-p} = u \cdot n$.

Drawing the observed pictures in Table 4-12.

Table 4-12　Data record for the measuremeant of Lissajous' picture

Signal from channel 2	About 25 Hz	About 50 Hz	About 75 Hz	About 100 Hz	About 200 Hz
Real value of signal from channel 1/Hz					
Picture					
f_y/f_x					
$f_y/$ Hz					

$\overline{f_y} = $ _____ Hz.

VI. Questions

(1) Please briefly describe the structure of oscilloscope and the function of each part.

(2) Try to describe the adjustment steps to obtain the following images on the oscilloscope screen:

i. a bright spot;

ii. a horizontal bright line;

iii. a vertical bright line;

iv. a 50 Hz sinusoidal waveform.

(3) Please briefly describes the reason why the oscilloscope can truly display the input signal waveform.

(4) According to the oscilloscope, how to measure the effective value and frequency of AC signal voltage?

(5) What are the conditions for the waveform stability? How to adjust the relevant knobs of oscilloscope to make the waveform stable?

(6) How to use Lissajous' picture to measure the frequency of sinusoidal signal?

Chapter 5

Normal experiment

Experiment 1   Measurement of electromotive force of power supply with linear DC potentiometer

I. Purpose

(1) To understand the working principle of potentiometer.

(2) To master the method of measurement electromotive force and potential difference of power supply with linear potentiometer.

II. Instruments

A linear potentiometer with eleven lines, galvanometer, rheostat, DC regulated power supply, switch with one pole, switch with double poles, a standard battery, two batteries to be measured.

III. Principle

1. The compensation principle

Figure 5-1 shows the schematic of voltage compensation circuit, in which E_0 is a compensation power, and E_x is an unknown voltage to be measured, G is a galvanometer. When we turn on the switch K, if $E_0 \neq E_x$, there will be a current pass through G. If we adjust the potential of the compensation power until the $E_0 = E_x$, the current flowing through the galvanometer G becomes zero. In this case, the pointer of the galvanometer points the zero line. This method is called compensation method.

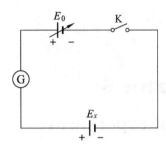

Figure 5-1   Circuit of voltage compensation principle

2. Working principle of linear potentiometer

Figure 5-2 shows the schematic of the working circuit of the linear potentiometer. We define the voltage between points C and D as U_{CD}, the resistance between C and D as R_{CD}. Firstly, the working circuit should have a constant working current. This process is called the standardization of working current (also known as potentiometer standardization), which can be achieved by using the standard batteries. Secondly, switch K_2 to connect the standard battery E_n into the circuit. The length of the resistance wire CD is selected appropriately, and the resistance between C and D is R_{CD}. Finally, adjust the resistance of the rheostat R_p, and make the current flowing through the galvanometer G be zero. In this case, the potential difference between C and D equals the electromotive force of the standard battery.

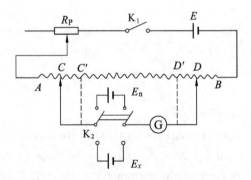

Figure 5-2   Circuit of linear potentiometer

We have

$$E_n = U_{CD} = IR_{CD} \propto l_n I \tag{5-1}$$

where I is a constant current, L_n is the length of resistance wire between C and D.

After the current standardized process, we can use the linear potentiometer to measure E_x. Turn the switch K_2 toward E_x, adjust the length of the resistance wire between C and D until the galvanometer's reading becomes zero again. In this situation, the length of resistance between C and D is changed to be l_x, we have

$$E_x = U'_{CD} = IR'_{CD} \propto I l_x \tag{5-2}$$

Comparing formula (5-1) and (5-2), we obtain

$$\frac{E_x}{E_n} = \frac{l_x}{l_n}, \text{ namely } E_x = \frac{l_x}{l_n} E_n \tag{5-3}$$

If we set $\dfrac{E_n}{l_n} = V_0$, then

$$E_x = V_0 l_x \tag{5-4}$$

In this experiment, we set $V_0 = 0.300\ 00$ V/m.

IV. Contents and steps

(1) Construct the circuit based on Figure 5-2. First, K_1 is turn off.

(2) Record the room temperature t, and obtain the $E_n(t)$ from table on the wall of the classroom.

(3) Calculate L_n using $l_n = \dfrac{E_n(t)}{0.300\ 00}$.

(4) Turn the switch K_2 toward E_n.

(5) Adjust the distance between C and D until the length is equal to L_n.

(6) Adjust the resistance of the rheostat R_p, until the pointer of the galvanometer points zero.

(7) Turn the switch K_2 toward E_x, adjust the distance between C and D until the galvanometer's reading becomes zero again. Record the length of the resistance wire between C and D as l_x.

(8) Repeat steps 4, 5, 6 and 7, measure the electromotive force values of two batteries (denoted as E_{x1} and E_{x2}), respectively. Then, connect the two batteries in series and measure the total electromotive force.

V. Data record and process

$t =$ _____ ℃, $E_n =$ _____ V,

$V_0 = 0.300\ 00$ V/m, $l_n = \dfrac{E_n}{V_0} =$ _____ m.

The measured values of l_x are filled in Table 5-1.

Table 5-1　Data record for the measurement of electromotive force of the batteries

	L_{x1}/mm	L_{x2}/mm	$\dfrac{(L_{x1}+L_{x2})}{2}$/mm	$E_x = \dfrac{L_{x1}+L_{x2}}{2}$/V
Battery 1				
Battery 2				
Series connection				

VI. Questions

(1) If the pointer of the galvanometer always deflected towards to the same side in the experiment and the galvanometer cannot be adjusted to zero, what are the possible reasons?

(2) When $V_0 = 0.300\ 00$ V/m, what is the measurement range of the linear potentiometer used in this experiment?

(3) Please briefly describe the compensation principle of potentiometer, and use this principle to measure resistance according to voltammetry, and draw a circuit diagram.

Experiment 2　Measurement of resistance using
Wheatstone bridge

I. Purpose

(1) To master the principle of resistance measurement with Wheatstone bridge.

(2) To learn to use the Wheatstone bridge to measure resistance correctly.

(3) To understand how to improve the sensitivity of the bridge.

(4) To learn the method to eliminate systematic errors.

II. Instruments

QJ23 single arm bridge, wiring board for bridge (including R_1, R_2, and four terminals), resistance box, galvanometer, regulated power supply, resistance to be measured, switch and wireway.

III. Principle

Bridge is a kind of instrument to measure resistance by comparing electric potential. It can be divided into two categories: direct current bridge and alternating current bridge. The direct current (DC) bridge could also be divided into single arm and double arm bridge. The single arm bridge is usually called Wheatstone bridge, which is mainly used to measure the resistances with their values between $1 \sim 10^6$ Ω. The double arm bridge is usually called Kelvin bridge, which is mainly used to measure the resistance with their values between $10^{-5} \sim 1$ Ω.

1. Principle for Wheatstone bridge

The Wheatstone bridge is consisted of four branches which are labeled as AD, DC, CB and BA, respectively. Each branch contains a resistance, as shown in Figure 5-3. We join up the B and D points and connect a galvanometer which looks like a bridge, connecting the lower ABC and upper ADC branches. The electric power supply and switch are connected between A and C points. When we turn on the switch, the current passes through each branch. If the electric potentials at B and D points are not equal, the current will flow through the galvanometer. We can adjust the resistances in each branch to make the bridge reach a equilibrium state in which the electric potentials at B and D points are equal and the pointer of the galvanometer points zero.

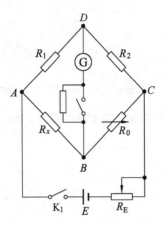

Figure 5-3   Schematic of Wheatstone bridge

Base on the voltage divide principle,

$$U_{DC} = U_{AC} \frac{R_2}{R_1 + R_2}$$

$$U_{BC} = U_{AC} \frac{R_0}{R_x + R_0} \qquad (5\text{-}5)$$

When the bridge reaches the equilibrium state, we have $U_{DC} = U_{BC}$, namely $\frac{R_2}{R_1 + R_2} = \frac{R_0}{R_x + R_0}$, from which we can deduce that

$$R_x = \frac{R_1}{R_2} \cdot R_0 \qquad (5\text{-}6)$$

In this experiment, R_1 and R_2 are constant resistances, and $\frac{R_1}{R_2}$ is defined as the ratio arm of the Wheatstone bridge, while R_0 is a rheostat. We can adjust the value of R_0 to make the bridge reaches the equilibrium state in which the galvanometer indicates a zero current. R_x will be calculated by using formula (5-6).

2. Sensitivity of Wheatstone bridge

The measurement accuracy of the Wheatstone bridge is determined by its sensitivity which is defined as

$$S = \frac{\delta n}{\delta R_0 / R_0} \qquad (5\text{-}7)$$

here δR_0 is a deviation of R_0, and δn is the grid number of the galvanometer's pointer deflected away from the zero-line induced by the deviation δR_0.

3. Correction of systematic error

If we exchange the position of R_1 and R_2 in Figure 5-3, the equilibrium state of the bridge will be broken. However, we can adjust R_0 to make the bridge reach a new equilibrium again. In this case, the value of R_0 will be changed to R_0'. For the two equilibrium states, according to formula (5-6), R_x will be expressed as $R_x = \frac{R_1}{R_2} R_0$ and $R_x = \frac{R_2}{R_1} R_0'$, respectively. Therefore, we can obtain that $R_x = \sqrt{R_0 \cdot R_0'}$, from which we can see that the systematic error arising from R_1 and R_2 is eliminated.

IV. Contents and steps

1. Measurement of resistance by using QJ23 single arm bridge

(1) Connect the unknown resistance R_{x1} and R_{x2} to the input terminals of QJ23 single arm bridge.

(2) Open the power of QJ23 single arm bridge.

(3) Adjust the zero-adjusting knob to make the pointer of the galvanometer point the zero line.

(4) Select the proper ratio arm based on the estimation of the resistance to be measured.

(5) Press the controlling button of the galvanometer, adjust until the galvanometer's pointer points the zero, then turn off the power.

(6) Calculate using formula (5-6).

(7) Record the measured data with connect R_{x1}, R_{x2}, series connection R_{x1} and R_{x2}, parallel connection R_{x1} and R_{x2}, respectively.

2. Measurement of sensitivity

(1) Change R_0 with a small value δR_0. As a result, the galvanometer's pointer would shift away from the zero line, the offset of pointer δn should be recorded.

(2) Sensitivity of Wheatstone bridge is calculated by formula (5-7).

3. Elimination of systematic error

(1) Constructing a Wheatstone bridge using wiring board base on Figure 5-3. Adjust the rheostat until the bridge reaches the equilibrium state, and record the value of R_0.

(2) Exchange the positions of R_1 and R_2, adjust the rheostat R_0 until the bridge reaches the equilibrium state again, record the value of rheostat R_0'.

V. Data record and process

(1) Measurement with QJ23 single arm bridge

Class of rheostat's accuracy $a = $ _____.

The measured values are filled in Table 5-2.

Table 5-2   Data record for the measurement of the ratio arm and R_0

The measured resistance	R_{x1}	R_{x2}	Series connection	Parallel connection
Ratio arm				
R_0/Ω				
Measurement value/Ω				

(2) Sensitivity of QJ23 single arm bridge

$R_0 =$ _____ Ω (galvanometer' pointer indicates zero),

$R_0' =$ _____ Ω (galvanometer's pointer shifts n div),

$\delta n =$ _____ div,

$\delta R_0 = |R_0 - R_0'| =$ _____ Ω,

$S = \dfrac{\delta n}{\delta R_0 / R_0} =$ _____.

(3) Measurement using combination bridge

Class of rheostat's accuracy $a =$ _____.

The measured values are filled in Table 5-3.

Table 5-3   Data record for the measurement of the resistances by using combination bridge

The measured resistance	R_{x1}	R_{x2}	Series connection	Parallel connection
Ratio arm				
R_0/Ω				
Measurement value /Ω				

$R_0 =$ _____ Ω (before exchange R_1 and R_2),

$R_0' =$ _____ Ω (after exchange R_1 and R_2),

$R_x = \sqrt{R_0 \cdot R_0'} =$ _____ Ω,

$E = \dfrac{\Delta_{R_x}}{R_x} = \dfrac{\Delta_{R_0}}{\sqrt{2} R_0} = \dfrac{1}{\sqrt{2}} \times a\% =$ _____,

$\Delta_{R_x} = R_x \cdot E =$ _____ Ω,

$R_x = R_x \pm \Delta_{R_x} =$ _____ Ω.

VI. Questions

(1) Why should we make Wheatstone bridge have 4 digits during the measurement of the resistance?

(2) During the measurement of the resistance with a combined bridge, we adjust the resistance box R_0, and find that the galvanometer G always deviates to one side and cannot be adjusted to zero, what is the possible reason?

Experiment 3   Refitting and calibration of electricity meter

I. Purpose

(1) To measure the resistance of galvanometer using half deflection method.

(2) To refit the ammeter by using galvanometer.

II. Instruments

DC regulated power supply, 100 μA galvanometer, DC ammeter, rheostat box, switch, wires and DC micrometer.

III. Principle

1. Refitting ammeter

As shown in Figure 5-4, we set the measuring range of galvanometer G as I_g, and the inner resistance as R_g. Generally, to measure a large current, we connect a resistance R_s in parallel at both ends of the galvanometer G. In this case, the part of the current which is expressed as $(I-I_g)$ will flow through R_s. Therefore, the refitted ammeter with a large measuring range of $0 \sim I$ is composed of the galvanometer G and resistor R_s. By properly select the value of R_s, we can refit the ammeter with different measuring range.

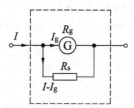

Figure 5-4   Schematic of galvanometer with a parallel connected resistance

Here, the maximum current allowed to pass through the galvanometer is only 100 μA. And the measuring range of the ammeter to be refitted is $0 \sim 5$ mA. When a current with a value of 5 mA is input into the refitted ammeter, the galvanometer will reach full scale deflection, namely, a current of 100 μA(I_g) passes through the galvanometer. In this situation, we have

$$R_s(I-I_g)=I_g \cdot R_g \qquad (5\text{-}8)$$

From formula (5-8), we finally get

$$R_s=\frac{I_g}{I-I_g}R_g \qquad (5\text{-}9)$$

By substituting the values of I_g, I and R_g into formula (5-9), we may obtain the value of R_s. Therefore, the measurement of the inner resistance of the galvanometer G is the key step in this experiment, which we will introduce in the following part.

2. Measure the inner resistance of the galvanometer by half deflection method

As shown in Figure 5-5, we can adjust the resistance R_1 or R_2 until the current passing through the standard micrometer equals I_g. Then, we adjust the rheostat R_0 until the pointer of the galvanometer reaches the half scale deflection. Repeat the above steps, until the micrometer shows a current with value of I_g and galvanometer shows a current with value of $1/2I_g$. In this case, $R_g = R_0$. This method is called half deflection method.

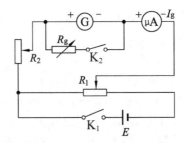

Figure 5-5   Circuit for the half deflection method to measure inner resistance of galvanometer

3. Meter error and calibration

In normal conditions, the possible maximum error of the electric meter included in the reading data is $\sigma_m = A_m \cdot K\%$, where σ_m is the maximum instrument error, K is the accuracy level of the meter, and A_m is the measuring range of the meter. In order to reduce the error of the electric meter, we may not use the accuracy level of the meter as the last judgement basis to determine the error. Usually, we read the indicating values I_x from the refitted ammeter, and compare them with the corresponding vales I_s from the standard ammeter. Then calculate the calibration values by $\delta I_x = I_x - I_s$, and plot the calibration curve, as shown in Figure 5-6. The curve is polyline. When we use the refitted electric meter, we can correct the reading data according to the calibration curve. It could be seen that the accuracy level of the refitted meter is determined by the calibration curve.

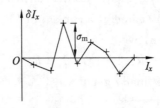

Figure 5-6   Schematic of calibration curve

IV. Contents and steps

1. Refitting the ammeter

(1) Construct a circuit based on Figure 5-5. Adjust the resistance R_1 or R_2 until the pointer of galvanometer points the 100th scale line, and read the data from the standard micrometer, record the value of I_g.

(2) Using the half deflection method to measure the inner resistance of the galvanometer, then using formula (5-9) calculate the value of R_s.

(3) Construct a circuit based on Figure 5-7. In this case, if the pointer of the galvanometer points the 10th scale line, it indicates that there is a current with value of 0.5 mA passing through the galvanometer. If the pointer of galvanometer points the 100th scale line, it indicates that there is a current with value of 5 mA flowing through the galvanometer.

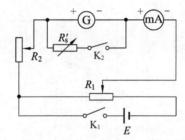

Figure 5-7   Circuit of the calibration of the refitted ammeter

2. Calibration of the refitted ammeter

(1) Adjust R_1 or R_2, make the pointer of the galvanometer points zero, record the reading from the standard ammeter. Then, increase the current with a step of 0.05 mA by adjusting the resistance R_1 or R_2 and recording the corresponding data from the standard ammeter, until the current reading from the refitted ammeter reaches 5 mA.

(2) Adjust R_1 or R_2, make the pointer of the galvanometer points 5 mA, record the reading from the standard ammeter. Then, decrease the current with a step of 0.05 mA by adjusting the resistance R_1 or R_2 and recording the corresponding data from the standard ammeter, until the current reading from the refitted ammeter reaches zero.

V. Data record and process

$I_g =$ _____ μA, $R_g =$ _____ Ω, $R_s =$ _____ Ω.

The measured values of the currents are filled in Table 5-4.

Table 5-4　Data record for refitting ammeter

Galvanometer's readings I_x/mA	0.00	0.50	1.00	1.50	2.00	2.50	3.00	3.50	4.00	4.50	5.00
Standard ammeter's readings I_1/mA(increase)											
Standard ammeter's readings I_2/mA(decrease)											
Average I_s/mA											
$\delta I_x = I_x - I_s$/mA											

$$I_s = \frac{I_1 + I_2}{I} = \underline{\hspace{2cm}} \text{ mA}.$$

Draw the picture of I_x-δI_x below:

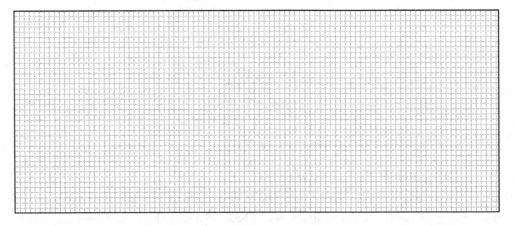

VI. Exercises

(1) Draw the circuit diagram of converting the galvanometer into an ammeter with two ranges (the ammeter has three terminals in total).

(2) Draw the circuit diagram of converting the galvanometer into a voltmeter with two ranges (the voltmeter has three terminals in total).

(3) Draw the circuit diagram of refitting the meter head into a dual-functionalities meter that can be used as both an ammeter and a voltmeter (the dual-functionalities meter has three terminals in total).

Experiment 4　Michelson interferometer

I. Purpose

(1) To master the adjustment and usage of Michelson interferometer.

(2) To observe the equal inclination and equal thickness interference patterns created by Michelson interferometer, and deepen the understanding of the characteristics of various interference patterns.

(3) To master the method of measuring laser wavelength with Michelson interferometer.

II. Instruments

WSM-100 Michelson interferometer, He-Ne laser, sodium lamp, white screen.

III. Principle

Michelson interferometer is a precision optical instrument designed and manufactured by American physicist Michelson and Morley in 1883, which would be used to study the "Ether" drift experiment. It can measure the wavelength of light with highly accuracy. People developed a variety of special interferometers based on the principle of the Michelson interferometer, which were widely used in modern physics and modern metrology technology.

Figure 5-8 shows the light path in the Michelson interferometer, in which M_1 and M_2 is a pair of plane mirrors, M_1 is fixed, M_2 can move along the guide track, G_1 and G_2 are a pair of parallel glass plates with the same thickness and refractive index. The angle between the $G_1(M_1)$ and $G_2(M_2)$ are 45°. The back surface of the plate G_1 is attached with a semi-reflective and semi-transmissive film A, on which the incident light is divided into reflected and transmitted lights with almost equal light intensity. For the transmitted light (1), it passes through G_2 and propagates to M_1. After the reflection by M_1, the light will pass through G_2 again, and be reflected by the semi-reflective film A on G_1. Finally, the light comes to E. For the reflected light (2), it propagates to M_2, and come back due to the reflection at M_2. Then the reflected light passes through G_1 and arrives at E. Note that, because the light (2) passes through G_1 three times, and the light (1) passes through G_1 only once but passes through the compensation plate G_2 twice, the paths of the lights (1) and (2) in the glass plates equal with each other. Therefore, when we calculate the light path difference of the two beams, we only consider the light paths in air.

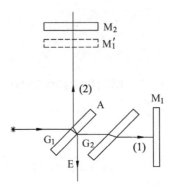

Figure 5-8   Light path in Michelson interferometer

As shown in Figure 5-9a, when the laser passes through a short focal length lens L, it becomes a point light source (marked as S) which emit the light to the Michelson interferometer. After the reflections on mirrors M_1 and M_2, S_1' and S_2' are equivalent to two coherent beams. Here, S' is the image of light source of S formed by the semi-reflecting surface A. S_1' and S_2' are the images of S' in mirrors M_1 and M_2, respectively. If we place a screen in the overlapping area of light emitted by two equivalent light sources S_1' and S_2', we can observe the interference patterns. This interference is called non-localized interference. If M_2 is strictly parallel to M_1' and the observation screen is placed on the line perpendicular to the connecting line between S_1' and S_2', a group of bright and dark concentric interference fringes can be observed. The center of the circle is located at the intersection of the screen and line $S_1' S_2'$. As seen in Figure 5-9b, the light path difference of two coherent lights at P_0 is $\Delta = 2d$, and the light path difference of any other points P' or P'' on the screen is approximately

$$\Delta = 2d \cos \varphi \tag{5-10}$$

According to the interference law, when $2d \cos \varphi = k\lambda$, the corresponding fringes are bright, however, the fringes are dark for $2d \cos \varphi = (2k+1)\lambda/2$, where k is the integer and called interference level. It can be seen from Figure 5-9b that, the interference patterns of concentric circles with P_0 as the center is induced by the light with different inclination angles. Note that, the same level interference fringe is generated by the incident light with the same inclination angle. Therefore, they are called equal inclination interference fringes.

Based on formula (5-10), the light path difference will reach the maximum when $\varphi = 0$, namely the level of the interference fringe at the center P_0 is the highest, which will decrease with the deviation from the center P_0.

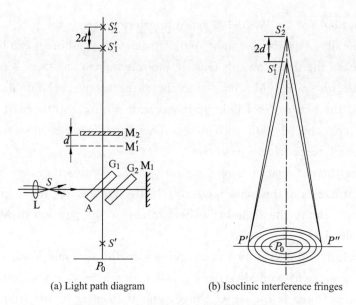

(a) Light path diagram   (b) Isoclinic interference fringes

Figure 5-9   Interference light path diagram of
point light source and its isoclinic interference fringes

With the increase of the distance d, the level of the center interference circle increases, and the fringes move outward along the radius. In experiment, we can observe that the interference ring emerges from the center and diffuses towards surrounding. In contrast, the interference ring shrinks to the center when d decreases.

According to the condition of bright fringe, $\Delta=2d=k\lambda$ when the center of the interference ring is bright. If the mirror M_2 moves(chang d), the level of the fringe at the center will be changed. If d changes per $\lambda/2$, there will be one interference ring emerges from the center or shrinks to the center. Here, we define the moving distance of M_2 as Δd, and the number of the interference rings emerging or shrinking in the center as N, the relationship between them could be written as

$$\Delta d = N\,\frac{\lambda}{2} \tag{5-11}$$

In experiment, the positions of the M_2 before and after its movement are recorded as d_1 and d_2, formula (5-11) could be rewritten as

$$\lambda = \frac{2\Delta d}{N} = \frac{2(d_1-d_2)}{N} \tag{5-12}$$

According to formula (5-12), we can calculate the wavelength of the light.

IV. Contents and steps

1. Adjustment of interferometer

(1) Rotate the scale handwheel to move M_2 until it arrives at the position

which is about 30 mm for WSM-100 typed interferometer.

(2) Generally, two rows of light spots (image of light source) can be observed in the view from the front towards G_1, we carefully adjust the three screws behind the fixed reflector mirror M_1 (note that the three screws behind M_2 cannot be rotated) until the two rows of light spots coincide with each other. At this time, M_2 and M_1' are parallel with each other. Therefore, we can observe the equal inclination interference fringes on screen.

(3) Place the observation screen in front of G_1. At this time, we can see the interference fringes on the screen clearly. If the fringes do not appear or are unclear, slowly rotate the scale handwheel to adjust the position of M_2 until the fringes become clear.

(4) Carefully adjust the two micro-screws on the side and lower ends of the reflector M_1 to make M_2 and M_1' strictly parallel. In this case, it should be seen that the interference fringe is a circular fringe, and the center of the fringe is at the center of the field of view, which is called the isoclinic interference.

2. Observe the interference pattern

(1) Observe the isoclinic interference pattern. After obtaining the isoclinic interference pattern, slowly rotate the scale handwheel to make the intervald between M_2 and M_1' gradually decrease to be zero. Continuously ously move M_2 in the original direction to make d increase. Observe the change of isoclinic interference pattern and record it.

(2) Observe the equal thickness interference pattern. After obtaining the equal thickness interference pattern, rotate the scale handwheel to make the interval d between M_2 and M_1' gradually decrease to be zero, and then increase from zero. Observe the change of the equal thickness interference pattern and record it.

3. Measure the laser wavelength

(1) Calibrate the zero position of the handwheel before the measurement. First, rotate the fine-tuning handwheel counterclockwise to align the reading guide line with the zero-scale line, and then rotate the coarse-tuning handwheel counterclockwise to align the reading guide line with a scale line. Of course, you can also rotate the handwheel clockwise to calibrate the zero position. However, it should be noted that, the handwheel rotation during the measurement should be consistent with that during calibration.

(2) Turn the fine-tuning handwheel in the original direction (change the value of d), and we can see that one interference ring emerges (or shrinks) from the center. When the center of the interference ring is the brightest, write down the position reading d_1 of the movable mirror, and then continue to rotate the fine-

tuning handwheel slowly. When the number of rings emergs (or shrinks) $N=50$, record the position reading d_2 of the movable mirror, and repeat measurement for many times. Calculate the wavelength from formula (5-12). Compare with the standard value ($\lambda_0=632.8$ nm), and calculate the relative uncertainty.

V. Data record and process

(1) Record the interference phenomena of equal inclination and equal thickness observed in the experiment. Draw the figure of the patterns of the interference fringe.

(2) Measure the wavelength of the laser.

The measured positions d are filled in Table 5-5.

Table 5-5　Data record for the measurement of the wavelength　　　　mm

Measuring times	d_i	$\Delta d_i = d_{i+5} - d_i$	$\overline{\Delta d} = \dfrac{\sum\limits_{i=1}^{5} \Delta d_j}{5}$	$\lambda = \dfrac{2\,\overline{\Delta d}}{250}$
1				
2				
3				
4				
5				
6				
7				
8				
9				
10				

$$\lambda = \frac{2\,\overline{\Delta d}}{5N} = \underline{\qquad}\ \text{mm},$$

$$E = \frac{|\lambda - \lambda_0|}{\lambda_0} = \underline{\qquad}\%.$$

VI. Questions

(1) What conditions do equal inclination interference fringes require? And what conditions do equal thickness interference fringes require?

(2) What is the difference between the isoclinic interference fringes produced by Michelson interferometer and Newton's ring?

(3) How to measure the wavelength difference $\Delta\lambda$ between two spectral lines of sodium light with Michelson interferometer? Try to explain its principle and steps.

Experiment 5　Newton's ring and wedge interference experiment

I. Purpose

(1) To observe the equal thickness interference phenomenon, and understand the characteristics of fluctuation of the light.

(2) To learn the measurement of the curvature radius, micro-thickness or diameter of the lens by interferometry.

(3) To master the principle and usage of reading microscope.

II. Instruments

Reading microscope, sodium lamp, Newton's ring instrument, and wedge instrument.

III. Principle

1. Newton's ring

As shown in Figure 5-10, a flat convex lens A with a large curvature radius is placed on the flat glass plate B. The device is called Newton's ring instrument. When the convex surface of the lens and the flat glass plate contact with each other, an air film layer is formed between A and B, whose thickness gradually increases from the central point to the edge. The thicknesses of the air film at the positions with the same distance from the center point are the same. For the normal incidence of the light from the top, the reflected lights induced by the lower surface of the convex lens and upper surface of the glass plate will interfere with each other. The interference fringe is the track of the equal thickness point of the air film. Therefore, this interference phenomenon is called the equal thickness interference. And the interference pattern exhibits a series of concentric circles centered on the contact point, which is named as Newton' ring.

We define that the wavelength of the incident light is λ, the thickness of the air film at a position with the distance r from the center is d, and the curvature radius of the convex lens is R. For the case of normal incident, the light path difference of the reflected lights on the upper and lower surfaces is

$$\delta = 2d + \frac{\lambda}{2} \tag{5-13}$$

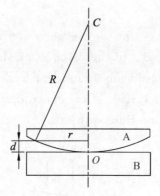

Figure 5-10   Schematic of Newton's ring

where $\dfrac{\lambda}{2}$ is the additional optical path difference induced by the half-wave loss.

According to the geometric relation in Figure 5-10, we have

$$R^2 = r^2 + (R-d)^2 = R^2 - 2Rd + d^2 + r^2 \tag{5-14}$$

Because comparing with the R, d is very small, we can ignore the term d^2. Therefore, we obtain that

$$d = \frac{r^2}{2R} \tag{5-15}$$

We set the thickness of the air film corresponding to the k-th dark fringe as d_k. Based on formula (5-13), we have

$$2d_k + \frac{\lambda}{2} = (2k+1)\frac{\lambda}{2} \tag{5-16}$$

where k is the level of interference fringes, namely $k = 0, 1, 2, 3, \cdots$.

Combined with formulas (5-15) and (5-16), we get

$$r_k = \sqrt{kR\lambda} \tag{5-17}$$

Therefore, if we know the wavelength and the radius of the k-th dark fringe, we can calculate the curvature radius R of the convex lens.

However, in practice, the observed ring number is inconsistent with the level of the fringe, due to the elastic deformation occurred between the lens and glass. Besides, it is very difficult to determine the center of the circle fringes and measure their radius accurately. Note that, there always exists one level difference between the adjacent two dark fringes (or bright fringes), therefore, we measure the diameter difference between the two dark rings to eliminate the error induced by the above problems. We have

$$R = \frac{D_m^2 - D_n^2}{4(m-n)\lambda} \tag{5-18}$$

where D_m and D_n are the diameters of the m-th and n-th dark fringes, respectively.

2. Air wedge

As shown in Figure 5-11, the air wedge is constructed by two glass plates stacked at one end and inserted a filament (or a thin sheet) with diameter D at the other end. If the light is perpendicularly incident from the top, the two beams of light reflected from the lower surface of the upper glass plate and the upper surface of the lower glass plate will interfere with each other.

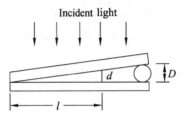

Figure 5-11   Schematic of air wedge

Similar with the Newton's ring, the thickness of the air film between two glass plates corresponding to the k-th dark fringe is given by

$$d_k = k \frac{\lambda}{2} \tag{5-19}$$

The interference fringes formed by the air wedge are a series of fringes parallel to the edge of the wedge.

However, in practice, it is difficult to determine the value of k. To overcome the problem, we measure the length L_x contains x interference fringes, such as $x = 30$. So, we can calculate the number of interference fringes per unit length $n = \frac{x}{L_x} = \frac{30}{L_x}$. If the distance between the edge of the wedge and the filament is L, the diameter of the filament is

$$D = n \cdot L \cdot \frac{\lambda}{2} = 15\lambda \frac{L}{L_x} \tag{5-20}$$

IV. Contents and steps

1. Measure the curvature radius of lens with Newton's ring

(1) Observe the Newton' ring instrument by using the sodium lamp.

(2) Place the Newton's ring instrument on the stage and make it face to the objective lens of the reading microscope. Adjust the focusing handwheel to make the microscope move down slowly until the lower end of the flat glass is close to the upper surface of the Newton's ring instrument.

(3) Change the relative position between the microscope and the light source, adjust the inclination angle of the flat glass to make the field of view in the eye lens bright. Rotate the eyepiece until the cross wire on the dividing plate can be seen

clearly. Loosen the eyepiece fixing screw, rotate the whole eyepiece until the vertical line is perpendicular to the main ruler of the microscope, and then fix the screw.

(4) Rotate the focusing handwheel, and make the microscope rises slowly, therefore, the microscope can focus on the interference ring and a clear Newton's ring can be observed. At this time, to make the center of the interference fringes roughly coincide with the intersection of the cross wire by slowly translating the Newton's ring instrument.

(5) Rotate the micrometer drum to move the cross-wire intersection from the center of the rings to left side. In order to eliminate the systematic error caused by the gap between the screw rod and the nut, we make the cross-wire intersection exceed the 30th circles of the dark ring, and then rotate the micrometer drum in reversed direction, make the vertical line of the cross-wire tangent to the middle of the 30th dark ring, and record the reading data. Rotate the micrometer drum continuously to make the vertical line of the cross hair tangent to the middle of the 29th, 28th, 27th, 26th, ⋯, 11th dark rings in turn, and record the readings. Then, rotate the micrometer drum to make the intersection of the cross wire cross the center of the Newton's ring to the right side. Move and record the readings of the 11th, 12th, 13th,⋯,30th dark rings in turn.

(6) Figure out the difference between the readings of the same dark ring on the left and right sides respectively, and calculate the diameter of each dark ring, then calculate the square value of each diameter. In order to make full use of all the measured data and improve the accuracy of measurement, the data is processed by successive differential method. Combine the square of the diameters of two dark rings separated by ten steps (the 30th and 20th rings, 29th and 19th rings, 28th and 18th rings, ..., 21th and 11th rings), in each group $m-n=10$. Calculate ten group of $D_m^2-D_n^2$, and take the average of $D_m^2-D_n^2$.

(7) Calculate the curvature radius of the lens by using $R=\dfrac{D_m^2-D_n^2}{4(m-n)\lambda}$.

2. Measure the diameter of metal wire with wedge interference fringes

(1) Place the wedge instrument (the metal wire has been clamped at one end of two glasses) on the platform to observe the interference fringes from the eyepiece. If the fringes are not clear, adjust the focusing handwheel and move the microscope tube up and down until clear interference fringes are observed from the eyepiece.

(2) Measure the length of 30 fringes in three different places, and then substitute the average value into formula (5-20) to calculate D.

V. Data record and process

1. Measurement of curvature radius of lens

The measured values are filled in Table 5-6.

Table 5-6　Data record for the measurement of rdius R of the lens

$\lambda = 589.3$ nm

Ordinal Direction	30	29	28	27	26	25	24	23	22	21
Left/mm										
Right/mm										
Ordinal Direction	20	19	18	17	16	15	14	13	12	11
Left/mm										
Right/mm										
$D_m^2 - D_n^2/\text{mm}^2$										

Calculate curvature radius R of the lens by using successive differential method.

2. Measurement of the diameter of the metal wire

The measured values are filled in Table 5-7.

Table 5-7　Data for the measurement of diameter D of the metal wire

$\lambda = 589.3$ nm　　$L = \underline{\hspace{2cm}}$ mm

L_1/mm			
L_2/mm			
L_x/mm			

Calculate the diameter of the metal wire D according to formula (5-20).

VI. Questions

(1) Why not use formula $r_k^2 = kR\lambda$ to calculate R?

(2) In Newton's ring experiment, if chord length is measured instead of diameter, what will the experimental results be influenced?

(3) Can the refractive index of liquid be measured by the interference fringe produced by the liquid wedge?

Experiment 6   Measurement of magnetic field in electrified solenoid with Hall sensor

I. Purpose

(1) To learn the relationship between the excitation current in solenoid and the output of Hall sensor.

(2) To know the characteristics and applications of Hall sensor.

(3) To understand the relationship between the magnetic intensity and the position of inner solenoid.

II. Instruments

Solenoidal magnetic field measuring instrument including a solenoid with a long rod, which is connected with a Hall sensor on its top, DC regulated power supply.

III. Principle

1. Magnetic induction intensity of long straignt current carrying solenoid

As shown in Figure 5-12, the induced magnetic intensity at O point locating at the central axis of the solenoid is given by

$$B_O = -\frac{\mu_0 N I_M}{\sqrt{L^2 + 4R^2}} \tag{5-21}$$

where L is the effective length of the solenoid, I_M is the excitation current, R is the radius of solenoid, N is the turn number of the solenoid, and μ_0 is the permeability in vacuum.

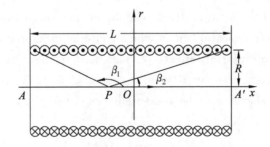

Figure 5-12   Sectional view of solenoid

2. Hall effect

The Hall effect is a well-known physical phenomenon, which arises from the deflection of the electrons propagating in a magnetic field induced by the Lorenz

College Physics Experiment

force.

As shown in Figure 5-13, a current I_C passes through the hall element, the electrons in the hall element will move with the current, and the magnetic field B is perpendicular to the direction of the current. Due to the effect of Lorentz force f_m, the electrons will be deflected and continuously gather towards the side D' of the hall element, and the other side D will carry equal positive charges. In the case, an additional electric field E_H is established between D and D', which is called hall electric field. Therefore, the electrons will be affected by the electric field force f_e induced by E_H. Note that, the electric field force is opposite with the Lorentz force, and will increase with the continuous accumulation of electrons at the D and D' sides. At the beginning, $f_e < f_m$, the electrons still are deflected and accumulated at the D and D' sides, until $f_e = f_m$, the charges at the D and D' sides no longer increase, forming a stable electric field. At this time, the voltage between the D and D' sides reaches a stable value, which is called the Hall voltage.

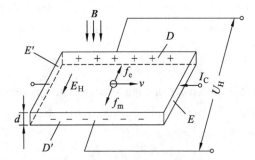

Figure 5-13　Schematic of Hall effect

The Hall voltage U_H, Hall sensitivity K and magnetic intensity B have a relationship as follows:

$$U_H = KB \tag{5-22}$$

Because the magnetic intensity B cannot be measured directly, we need to convert it into other physical quantity which could be measured directly and easily. Based on the principle of Hall effect, we just need to measure Hall voltage, and calculate the magnetic intensity B using formula (5-22). At the same time, the Hall sensitivity K also has to be measured. Combined the formulas (5-21) and (5-22), we may know that

$$K = \frac{\Delta U_H}{\Delta B} = -\frac{\Delta U_H}{\Delta I_M} \frac{\sqrt{L^2 + 4R^2}}{\mu_0 N} \tag{5-23}$$

IV. Contents and Steps

(1) Standardize the working current, set the working voltage as 2.500 V.

(2) Initialize the working current, set the inital current as zero.

90

(3) Insert the log rod with the Hall sensor into the solenoid, then adjust the insertion depth of the rod until the reading shows that $x=15.00$ cm.

(4) Set the excitation current I_M as 0, read the voltmeter and record the data. Increase the excitation current with a step of 50 mA, read and record the corresponding data, until the excitation current reaches 500 mA.

(5) Set the excitation current as 250 mA, and keep the current unchanged during the whole experiment. Firstly, set the position of the Hall sensor located inner the solenoid as $x=0$ cm, read the Hall voltage and record the data. Then change the sensor's position to $x=0.5$ cm, read the data and record it. Increase the position value x, and record the corresponding Hall voltage until $x=30.0$ cm. Finally, plot the figure of B-x.

V. Data record and Process

1. Measure ment of sensor's sensitivity

The measured values of the voltage are filled in Table 5-8.

Table 5-8   Data record for the measurement of the sensitivity of the sensor

$x=15$ cm

I_M/mA	0	50	100	150	200	250	300	350	400	450	500
U_H/mv											

The slope $\dfrac{\Delta U_H}{\Delta I_M}=$ _____ ,

$L=260$ mm, $N=300$, $\overline{D}=2R=35.0$ mm, $\mu_0=4\pi\times10^{-7}\mathrm{N/A^2}$.

Using formula $K=\dfrac{\Delta U_H}{\Delta I_M}=-\dfrac{\Delta U_H}{\Delta I_M}\dfrac{\sqrt{L^2+4R^2}}{\mu_0 N}$, calculate the $K=$ _____ .

2. The distribution of magnetic field

The measured U_H is filled in Table 5-9.

Table 5-9   Data record for the measurement of the magnetic field

$I_M=250$ mA

x/cm	0.00	0.50	1.00	1.50	2.00	3.00	4.00	5.00	6.00	7.00	8.00	9.00
U_H/mV												
$B=\dfrac{U_H}{K}$/mT												
x/cm	10.00	11.00	12.00	13.00	14.00	15.00	16.00	17.00	18.00	19.00	20.00	21.00
U_H/mV												
$B=\dfrac{U_H}{K}$/mT												

Continued

x/cm	22.00	23.00	24.00	25.00	26.00	27.00	28.00	28.50	29.00	29.50	30.00	
U_H/mV												
$B = \dfrac{U_H}{K}/\text{mT}$												

Draw the picture of B-x.

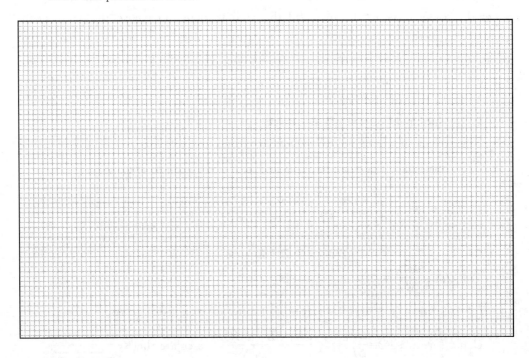

VI. Questions

(1) What is the Hall effect? What are the applications of Hall sensor in scientific research?

(2) What will happen if the number of turns per unit length on both sides of the solenoid is different or the winding is uneven?

Experiment 7　Measurement of sound speed

I. Purpose

(1) Try to measure the sound speed by phase-comparison method.

(2) Deepen the understanding of theoretical knowledge of standing wave and vibration synthesis.

(3) To cultivate the ability of comprehensive operation of instruments.

II. Instruments

Dual trace oscilloscope, sound velocity meter, and coaxial cable.

III. Principle

Sound wave is a kind of mechanical wave, which can travel in air and other elastic medium. The sound wave with the frequency in the range of $20 \sim 20\,000$ Hz is called hearing sound, and those below and above the range are called infrasound and ultrasonic, respectively. The sound has a great of applications in the engineering area. For example, the ultrasonic has an excellent ability to pass through the solid and liquid objects.

The ultrasonic has a shorter wave length, it can keep the propagation direction more easily than the other sound wave. Here, we adopt the ultrasonic to perform the experiment. The relationship of sound speed v, frequency f and wave length λ is given as

$$v = f \cdot \lambda \qquad (5\text{-}24)$$

1. Piezoelectric transducer

Piezoelectric transducer is made of piezoelectric ceramics, which can be elongated or compressed with the variation of the external electric field at the same frequency. We input the initial electric signal produced by signal generator into the piezoelectric transducer. The piezoelectric transducer could convert the initial electric signal to ultrasonic signal which would propagate a distance in air and finally reach the receiving end. We also place the other piezoelectric transducer at the receiving end which can convert the ultrasonic signal into the electric signal.

2. Phase-comparison method

We defined the ultrasonic wave at the transducer as

$$x = A_1 \cos(\omega t + \varphi_1) \qquad (5\text{-}25)$$

where the A_1 is the amplitude, ω is the angular frequency, and φ_1 is the initial phase. We divide the initial electric signal into two parts, one of which is input into channel 1 of the oscilloscope directly as the reference signal, and the other one is converted into the ultrasonic signal by the transducer at the emitting end. The ultrasonic propagates in air and is received by the transducer at the receiving end. We input it into channel 2 of the oscilloscope. Note that, the phase of the ultrasonic is delayed after its propagation in air with a distance L, which is changed to be φ_2. The received signal could be expressed as

$$y = A_2 \cos(\omega t + \varphi_2) \qquad (5\text{-}26)$$

Comparing formula (5-25) and (5-26), we will obtain that

$$\frac{x^2}{A_1^2} + \frac{y^2}{A_2^2} - \frac{2xy}{A_1 A_2} \cos(\varphi_2 - \varphi_1) = \sin^2(\varphi_2 - \varphi_1) \qquad (5\text{-}27)$$

We define the phase difference as $\Delta\varphi = \varphi_2 - \varphi_1$, which is calculated by

$$\Delta\varphi = \omega t = \omega \frac{L}{v} \qquad (5\text{-}28)$$

where the sound speed is $v = \lambda f$ and the angular frequency $\omega = 2\pi f$. Therefore, formula (5-28) can be rewritten as

$$\Delta\varphi = 2\pi L / \lambda \qquad (5\text{-}29)$$

If we synthesis the two signals by using the oscilloscope, Lissajous pictures are observed on the screen. As shown in Figure 5-14, when $\Delta\varphi = 0$, the Lissajous picture is a line with positive slope. And when $\Delta\varphi = \pi$, the slope of the line becomes negative. The Lissajous picture also could be ellipse when $\Delta\varphi = \pi/2$. Therefore, we can determine the phase difference according to the Lissajous picture.

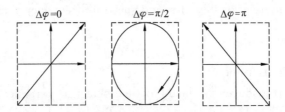

Figure 5-14　Lissajous pictures at different phase difference

As shown in Figure 5-15, the transducer S_1 is fixed and S_2 could move along the scale bar. We can adjust the phase difference $\Delta\varphi$ by moving the transducer S_2. If the Lissajous picture observed on the screen is changed from the line with positive slope to that with negative slope, we can determine that $\Delta\varphi = \pi$. According to formula (5-29), we can get $L = \frac{\lambda}{2}$. That is to say, when the phase varies a value of

π, the distance between S_1 and S_2 would change a half of wave length, namely $\frac{\lambda}{2}$.

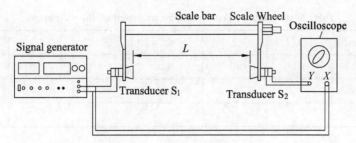

Figure 5-15   Schematic of test circuit

IV. Contents and steps

1. Preheating

(1) Open the power, and wait for several minutes.

(2) Select channel 1, adjust the t/div and V/div knobs, until we could see the initial wave clearly on screen.

(3) Select channel 2, redo the operation of step (2).

2. Measuring the length and speed of sound wave

(1) Select channel 2, adjust the frequency knob on signal generator until the amplitude of the signal on screen becomes the largest, and record the value of frequency. This frequency is so called piezoelectric transducer's resonance frequency. Working at this frequency, the signal received by piezoelectric transducer is the most clear.

(2) Adjust the position of S_2 by rotating the scale wheel on the right side of the scale bar, until the distance between S_1 and S_2 is about 10 mm.

(3) Rotate the scale wheel in counter clock wise, let S_2 moves away from S_1 until the Lissajous picture becomes a line with negative slope. Record the position of S_2 as l_1. Then, rotate the scale wheel again, making S_2 continue to move away from S_1 until the Lissajous picture becomes a line with opposite slope. Record the new position of S_2 as l_2.

(4) Redo the operation of step (3), until twelve groups of data have been measured. Using the twelve groups of data, calculate the length and speed of sound wave.

V. Data record and process

Room temperature $t=$_____℃, ultrasonic frequency $f=$_____ kHz.

$$L = \frac{1}{12}\sum_{i=1}^{12} l_i$$

The measured positions are filled in Table 5-10.

Table 5-10   Data record for the measurement of the wavelength

Measuring times	1	2	3	4	5	6
l_i/mm						
Measuring times	7	8	9	10	11	12
l_i/mm						

VI. Questions

(1) How to adjust and judge whether the measurement system is in resonance state?

(2) Why do we keep the transducer parallel to the surface of the receiver during the experiment?

(3) Combined with the experimental results, please analyze the causes of errors.

(4) Can the propagation velocity of the ultrasonic wave in liquid and solid be measured by the method in this experiment? How to measure?

Experiment 8   Electro-optic effect of liquid crystal

I. Purpose

(1) To measure the electro-optic effect curve, and obtain the turn off voltage and the threshold voltage.

(2) To measure the contrast ratio of the liquid crystal displayer (LCD) in different visual angle.

II. Instrument

Photoelectric effect experiment instrument.

III. Principle

If we add a voltage U at the top and bottom ends of LCD, the transmittance will decrease. Because the external electrical field can change the arrangement of the liquid crystal molecules.

The initial structure of liquid crystal is equivalent to a spiral optical waveguide. As shown in Figure 5-16, polarizers P_1 and P_2 are placed at the top and bottom ends respectively. The polarization directions of P_1 and P_2 are orthogonal to each other. The natural light is incident from the top, and propagates in the liquid crystal. The polarization direction of the light will rotate along the spiral optical waveguide. When the light reaches the bottom end, its polarization direction is the same as that of P_2. Therefore, the light can pass through P_2. In this case, the path for the light is turn on. However, if we add a voltage U between the top and bottom of the LCD, the rod-shaped molecule will align the direction of the electric field. As a result, the LCD becomes a direct optical waveguide. The light passing through the waveguide may keep its polarization direction which is orthogonal to P_2. Therefore, the light cannot pass through the P_2. In this case, the path for light is turn off.

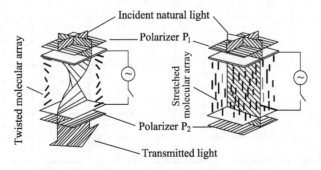

Figure 5-16   The schematic of the Electro-optic effect

Please note that:

(1) The turn off voltage U_2 is defined as the value of voltage at the two ends of LCD when transmittance reaches 10%.

(2) The threshold voltage U_1 is defined as the value of voltage at the two ends of LCD when transmittance reaches 90%.

(3) The contrast ratio is defined as $D = T(U_1)/T(U_2)$, where $T(U_1)$ is the transmittance with $U = U_1$, and $T(U_2)$ is the transmittance when $U = U_2$.

IV. Contents and steps

(1) Initialize the transmittance by holding the button to be pressed.

(2) Measure the relationship of the voltage and transmittance. Increase the voltage U and record the transmittance, until the transmittance approaches zero.

(3) Measure the ratio between the white and black states.

i. Set $U = U_2$, record the transmittance at different visual angles. The visual angle varies in the range of $-75° \sim 75°$.

ii. Set $U = U_1$, repeat step (1).

V. Data record and Process

1. Date for U-T

The measured transmittance is filled in Table 5-11.

Table 5-11 Data record for the measurement of transmittance

U/V	0	0.5	0.8	1.0	1.2	1.3	1.4	1.5	1.6	1.7	2.0	3.0	4.0	6.0
T														

2. Data for the measurement of contrast ratio

Table 5-12 is the data for the measurement of constrast ratio of the liguid crystal.

Table 5-12 Data record for calculation of the contrast ratio

Visual angle/(°)	0	5	10	15	20	25	30	35	40	45	50	55	60	65
$T(U_1)$														
$T(U_2)$														
D														
Visual angle/(°)	-5	-10	-15	-20	-25	-30	-35	-40	-45	-50	-55	-60	-65	
$T(U_1)$														
$T(U_2)$														
D														

Plot the figure of D and visual angle.

VI. Questions

(1) Please display the applications of the LCD.

(2) Can you discribe the principle of the LCD?

Experiment 9   Millikan oil-drop experiment

I. Purpose

(1) To determine the basic charge of the electron and verify the discontinuity of charge.

(2) To understand the basic principle and experimental method of measuring the charge of oil droplet by oil-drop meter.

II. Instruments

OM99 Millikan oil-drop meter (including oil drop box, oil drop lighting device, leveling system, microscope, power supply and other parts), monitor, oil spray and so on.

III. Principle

1. Determination of charge of oil droplet

Spray oil droplets using nebulizer, make the oil droplets enter the space between two horizontal parallel plates. The distance between the upper and lower plates is d. The ejected oil droplets are usually charged due to friction. Assuming that the mass of the oil droplet is m, the charge of the oil droplet is q, and the voltage between the two plates is U, the oil droplet will be subjected to gravity mg and static force $qE = q(U/d)$ at the same time between parallel plates. As shown in Figure 5-17, adjust the voltage U between the two plates to keep the oil droplets static, and then the two forces reach equilibrium, we obtain

$$mg = q\,\frac{U}{d} \tag{5-30}$$

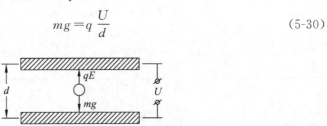

Figure 5-17   Diagram of force balance for oil droplet

In order to measure q, we need to measure m, d and U, respectively. The mass of the oil droplet is so small that it needs to be determined by the following special method.

2. Determination of mass of oil droplet

When the voltage applied to the parallel plate is zero, the oil drops fall faster under the force of gravity. However, due to the action of air resistance, the oil drops will move at a uniform speed after falling some distance (velocity is v). Then gravity and air resistance are in equilibrium (other forces can be ignored), as shown in Figure 5-18. According to Stokes's law of fluid mechanics, the air resistance of oil droplet is

$$f_r = 6\pi a \eta v = mg \tag{5-31}$$

where a is the radius of the oil drop (about 10^{-6} m), and η is the viscosity coefficient of the air.

Figure 5-18   Diagram of force analysis for oil droplet

Suppose that the density of oil is ρ, and the mass of oil droplet m can be expressed as

$$m = \frac{4}{3}\pi a^3 \rho \tag{5-32}$$

According to equations (5-31) and (5-32), the radius of the oil droplet is obtained as

$$a = \sqrt{\frac{9\eta v}{2\rho g}} \tag{5-33}$$

However, oil droplets are not rigid spheres, and their linearity can be compared with the average free path of gas molecules at room temperature (7×10^{-8}m), so Stokes's law is not strictly valid. By modifying η to η', we obtain

$$\eta' = \frac{\eta}{1+b/(pa)} \tag{5-34}$$

where b is the modified constant and p is the atmospheric pressure. Substituting equation (5-33) into equation (5-32), and replacing η in equation (5-33) with η', mass of an oil droplet is expressed by

$$m = \frac{4}{3}\pi \left[\frac{9}{2}\frac{\eta v}{\rho g} \cdot \frac{1}{1+b/(pa)} \right]^{3/2} \rho \tag{5-35}$$

3. Determination of v

Setting the distance through which the oil drops fall uniformly is l and the time required is t, the velocity of the oil droplets can be expressed as

$$v = \frac{l}{t} \tag{5-36}$$

Substituting equations (5-36) and (5-35) into (5-30), we get the expression for the charge of the oil droplet:

$$q = \frac{18\pi}{\sqrt{2\rho g}} \left[\frac{\eta l}{t(1+b/(pa))} \right]^{3/2} \frac{d}{U} \tag{5-37}$$

where $a = \sqrt{\dfrac{9\eta l}{2\rho g t}}$.

In this experiment, the density of oil is set as $\rho = 981$ kg $\cdot$ m^{-3}, acceleration of gravity is $g = 9.80$ m $\cdot$ s^{-2}, the viscosity coefficient of the air is $\eta = 1.83 \times 10^{-5}$ kg $\cdot$ m$^{-1} \cdot$ s^{-1}, $l = 1.50 \times 10^{-3}$ m, the modified constant is $b = 6.17 \times 10^{-6}$ m $\cdot$ cm(Hg), standard atmospheric pressure is $p = 76.0$ cm(Hg), and the distance between the upper and lower plates is $d = 5.00 \times 10^{-3}$ m.

Substitute the above data into equation (5-37), the charge of the oil droplet is obtained

$$q = \frac{0.927 \times 10^{-14}}{[t(1+0.022\ 6\sqrt{t})]^{3/2}} \cdot \frac{1}{U} \tag{5-38}$$

It can be observed in the experiment that the charge q carried by an oil droplet satisfies the following relation:

$$q = mg\frac{d}{U} = ne \tag{5-39}$$

where $n = \pm 1, \pm 2, \cdots$, and e is a constant. This means that the charge q is discontinuous and is an integer multiple of the minimum charge e, which is the charge of the electron:

$$e = \frac{q}{n} \tag{5-40}$$

where n is the number of electrons.

IV. Contents and steps

(1) Adjust the three leveling screws at the bottom of the oil dropper until the bubble of the level meter is in the central position, and the parallel plates are in the horizontal state.

(2) Adjust the balance voltage at about 200 V. Spray oil into the oil mist chamber and fine-tune the focusing hand wheel of the microscope, then a large number of oil droplet will appear on the display clearly.

(3) Measure the balance voltage. Look at a slow-moving oil droplet and adjust the balance voltage carefully until the droplet is stationary. Record the value of this balance voltage U.

(4) Superimpose the lifting voltage on the balanced voltage by the voltage

reversing switch to move the oil droplet to the first line on the display. Then switch the voltage to balance voltage so that the oil droplet remain stationary.

(5) Switch the voltage to 0 V. Press the timer button to start timing quickly when the oil droplet falls to the second line. When the oil droplet drops to the penultimate line press the timing button again to stop the timing, and then add the balanced voltage immediately to make the oil droplet stop. Record the time it takes for the oil droplet to descend the distance $l = 1.50$ mm.

(6) Superimpose the lifting voltage on the balanced voltage to make the oil droplet move to the appropriate position.

(7) Repeat the above steps for the same oil droplet for a total of five measurements.

(8) Measure five oil droplets in the same way.

V. Data record and process

(1) Record $\overline{U}$ and $\overline{t}$ of five oil droplets, and calculate the charge of each oil droplet according to equation (5-38).

(2) In order to prove the discontinuity of charge and that all charges are integral multiples of the basic charge e, and to find the value of e, we use the method of reverse verification for data processing. Use the accepted electron charge value e to divide the charge value q obtained from the above experiment and determine n (the nearest integer). The n is the number of fundamental charges that the oil droplets carry. And then use q/n to find the experimental value e.

(3) Calculate the average charge quantity $\overline{e}$ of five oil droplets and compare with the accepted value e_0, calculate the percentage error of experimental results.

$$e = \underline{\hspace{2cm}} \text{ C,}$$

$$E = \frac{|e - e_0|}{e_0} \times 100\% = \underline{\hspace{2cm}}.$$

VI. Questions

(1) When to add a lifting voltage, and when to remove it ?

(2) How to determine whether the oil droplet is in uniform motion?

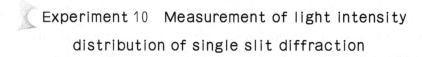

Experiment 10　Measurement of light intensity
distribution of single slit diffraction

I. Purpose

（1）To observe the phenomenon of Fraunhofer's diffraction.

（2）To master the method to measure light intensity distribution of Fraunhofer's diffraction, learn to measure the single slit's width using diffraction method.

II. Instruments

He-Ne Laser, single slit, light screen, light intensity distribution micrometer (consisted with silicon photocell, slit diaphragm and screw micrometer device), WJF digital galvanometer, optical bench, steel tape.

III. Principle

Diffraction is an important feature of light wave. Commonly, we classify diffraction into two types: one is Fresnel diffraction and another is Fraunhofer diffraction. For Fraunhofer diffraction, the distance between light source and diffraction screen is infinite, and distance between diffraction screen and receiving screen is also infinite. While for Fresnel diffraction, the distance between diffraction screen and receiving screen is finite.

1. The Huygens-Fresnel principle

During the light's propagation, the waves come from the same wave front are coherent waves, namely their initial phase are the same. When they meet with each other at a point in their propagation path, stacking pattern would appear which is called coherence stack. This is the Huygens-Fresnel principle.

2. Single slit diffraction

As shown in Figure 5-19, at P point the light's intensity could be expressed as

$$I = I_0 \frac{\sin^2 u}{u^2} \tag{5-41}$$

where $u = \dfrac{\pi a \sin \varphi}{\lambda}$, λ is the wave length, a is the width of single slit, φ is the diffraction angle, I_0 is the light intensity at the middle point P_0 of screen, L_2 is the lense.

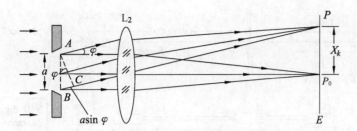

Figure 5-19　Schematic diagram of single slit's diffraction

Especially, when $\varphi=0$, namely $u=0$, then $I=I_0$. In this situation, the light intensity reaches maximum, and the bright stripe local at the middle of screen which is called the central bright stripe.

When $u=\pm\pi$, $I=0$. In this situation, the light's intensity is zero, and it shows as a dark stripe at each side of central bright stripe on the screen, shown in Figure 5-20.

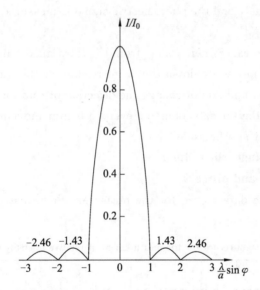

Figure 5-20　Light intensity distribution

If we define the distance between single slit and the receiver (silicon cell) as L, and the distance between central bright stripe and dark stripe on screen is X, we will have

$$a=\frac{\lambda L}{X} \tag{5-42}$$

IV. Contents and steps

The experimental setup is shown in Figure 5-21. Adjust the coaxial height of each element. The distance between the receiving screen and the single slit is greater than 1 m when the laser is illuminated vertically on a single slit plane.

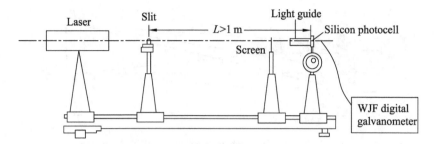

Figure 5-21   Diffraction intensity measurement system

1. Observe the single-slit diffraction

Adjust the width of the single slit carefully to observe the changes of the diffraction fringes, especially the intensity changes of the bright stripes at all levels.

2. Measure the relative intensity of the diffraction fringes

(1) Using a silicon photocell as a photodetector, the light signal is transformed into an electrical signal, and the photoelectric signal is measured with a WJF digital galvanometer.

(2) During the measurement, start from the third maximum center of one side of the diffraction stripe, write down the drum reading at this time, turn the drum in the same direction, and do not change the direction of rotation halfway. Read the reading of the WJF digital galvanometer every 0.5 mm movement until the third maximum center on the other side.

3. Measure the single slit width a

V. Data record and process

Table 5-13 is the data record for the measurement of distribution of the light intensity.

Table 5-13   Data record for the measurement of relative intensity of the light

time	1	2	3	4	5	6	7	8	9	10	11	12	13
x/mm													
$\dfrac{I}{I_0}$													
time													
x/mm													
$\dfrac{I}{I_0}$													

Continued

time	1	2	3	4	5	6	7	8	9	10	11	12	13
x/mm													
$\dfrac{I}{I_0}$													
time													
x/mm													
$\dfrac{I}{I_0}$													
time													
x/mm													
$\dfrac{I}{I_0}$													

Notes:

Draw the curve of current I with position x.

VI. Questions

(1) What effect5 does the change of the slit's width have on the diffraction fringes?

(2) What is the effect of the slot aperture width in front of the silicon photo-battery on the experimental results?

(3) If a transparent medium with a refractive index of n is filled in the spatial area from a single slit to the viewing screen, what is the difference filled with air?

(4) If a single slit Fraunhofer diffraction is observed with a white light source, what is the diffraction pattern?

Chapter 6

Innovative experiment

 Experiment 1　Study of heat engine working in air

I. Purpose

(1) To understand the principles of air heat engines and thermal cycle processes.

(2) To measure the thermal power conversion efficiency under different input power (temperature difference between hot and cold ends), and verify the carnot's theorem.

(3) To measure the relationship between the output power of the air heater and the load, and calculate the actual efficiency of the air heater.

II. Instruments

Air heater, heat source (optional electric heating or alcohol lamp heating), thermal engine tesfer, computer (or oscilloscope), thermometer.

III. Principle

The heat engine is a kind of instrument that convert the heat energy into mechanical energy. The first air heater was invented by Stirling in 1816, which becomes a basic experiment to study the second law of thermodynamics.

Commonly, the heat engine is consisted of several parts: high temperature room, low temperature room, working piston (WP), displacement piston (DP), cylinder, free wheel, heat source and so on.

As shown in Figure 6-1a, DP shifts towards left, and the air flows from low temperature room into high temperature room, the WP moves down. The air's temperature rises after entering the high temperature room, which induces the high pressure in high temperature room, the WP moves up, as shown in Figure 6-1b.

The air will push the DP to the right, the heat energy converts to free wheel's mechanical energy, and the air flows from high temperature to low temperature room, as shown in Figure 6-1c. The air's temperature drops in low temperature room, the WP moves down caused by the free wheel's inertia, the pressure in cylinder drops, as shown in Figure 6-1d. Thus, the heat engine finishes a cycle.

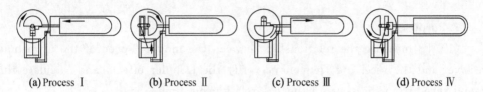

| (a) Process I | (b) Process II | (c) Process III | (d) Process IV |

Figure 6-1   Working process of heat engine

Base on Carnot's theorem, the efficiency of heat engines is

$$\eta = A/Q_1 = (T_H - T_L)/T_H = \frac{\Delta T}{T_H} \tag{6-1}$$

where A is the work outputted by heat engine, and Q_1 is the quantity of heat absorbed by heat engine.

IV. Contents and steps

Students are required to design experiment steps independently.

V. Data record and process

Students need to design data record sheets by themselves.

Experiment 2　Study of Doppler effect

I. Purpose

(1) To measure the relationship between the moving speed of the ultrasonic receiver and the receiving frequency, verify the Doppler effect, and calculate the sound speed from the slope of the f-v relationship line.

(2) To measure the velocity of multiple time points in the course of an object's motion by using Doppler effect.

(3) To validate Newton's second law, study the linear motion with uniform acceleration and measure the relationship between force, mass and acceleration.

II. Instrument

Doppler effect synthetic experimenter.

III. Principle

When there is relative motion between the wave source and the receiver, the phenomenon that the frequency of the received wave is different from that of the wave source is called Doppler effect. Doppler effect is widely used in scientific research, engineering technology, traffic management, medical diagnosis, etc. For example, the emission and absorption spectral lines of atoms, molecules and ions widen due to thermal motion, which is called Doppler broadening. In Astrophysics and controlled thermonuclear fusion experimental device, Doppler broadening of spectral line has become an important measurement and diagnosis method to analyze the physical state of stellar atmosphere and plasma. Radar system based on Doppler effect principle has been widely used to monitor the speed of moving targets such as missiles, satellites and vehicles. The principle of Doppler effect of electromagnetic wave (light wave) is the same as that of sound wave (ultrasound). This experiment can study not only the Doppler effect of ultrasound, but also the motion state of object by using the Doppler effect as a motion sensor.

According to the Doppler effect, the frequency of the received wave is

$$f = f_0 (u + v_1 \cos \alpha_1)/(u - v_2 \cos \alpha_2) \tag{6-2}$$

where f_0 is the frequency of the source, u is the speed of the ultrasound, v_1 is the speed of the receiver, v_2 is the speed of the source, α_1 is the angle between the connecting line of the sound source and the receiver and the moving direction of the receiver, and α_2 is the angle between the connecting line of the sound source and the receiver and the moving direction of the sound source.

IV. Contents and steps

Students are required to design experiment steps independently.

V. Data record and process

Students need to design data record sheets by themselves.

Experiment 3　Measurement of geomagnetic field

I. Purpose

(1) To understand the characteristics of magneto-resistive sensor, and master the calibration method of HMC1021Z magneto-resistive sensor.

(2) To measure the horizontal and vertical components of the geomagnetic field magnetic induction intensity and the magnetic inclination of the geomagnetic field.

II. Instruments

HMC1021Z magneto-resistive sensor, geomagnetic field measuring instrument, leveling device, heavy vertical line.

III. Principle

The earth is magnetic. There is a magnetic field around the earth, which is called geomagnetic field. The value of geomagnetic field is relatively small, the magnetic induction intensity is about the order of 10^{-5} T, and its intensity and direction vary with the position. However, in DC magnetic field measurement, especially weak magnetic field measurement, it is often necessary to know its value and try to eliminate its influence. As a natural magnetic source, geomagnetic field is widely used in military. It also has important applications in scientific research such as industry, medicine and exploration.

HMC1021Z magneto-resistive sensor is a one-dimensional magneto-resistive microcircuit integrated chip made of long and thin permalloy (iron nickel alloy). It often uses semiconductor technology to attach Fe Ni alloy films to silicon wafers. When a certain DC current is applied along the length direction of the iron nickel alloy belt and an external magnetic field is applied along the direction perpendicular to the current, the resistance value of the alloy belt will change greatly. The size and direction of the magnetic field can be measured by using the change of the resistance value of the alloy belt.

IV. Contents and steps

(1) Calibrate sensitivity of magneto-resistive sensor.

(2) Measure the magnetic induction intensity and magnetic inclination of the geomagnetic field.

Students are required to design experiment steps independently.

V. Data record and process

Students need to design data record sheets by themselves.

Experiment 4   Study of magnetic resistance

I. Purpose

(1) To observe the phenomenon of magnetic damping.

(2) To learn how to measure time by using a Hall sensor.

(3) To measure the velocity of magnetic slider sliding on the inclined plane of non-ferromagnetic body, and calculate the magnetic damping coefficient and sliding friction coefficient.

II. Instruments

Magnetic damping coefficient and sliding friction coefficient tester, balance.

III. Principle

Magnetic damping is an important concept in electromagnetism, and its mechanical effect is widely used in practice. The time measurement method of switching Hall sensor and single-chip microcomputer can change the timing measurement method from photosensitive sensor timing to magnetic sensor timing. Especially, it can still work through medium (non-magnetic medium), which makes its application prospect more extensive. It is a time measurement technology that is being popularized and applied on a large scale.

When a magnetic slider slides on the slope of a non-ferromagnetic good conductor with a uniform speed, the resistance of the slider arises from the sliding friction F_S and magneto-resistive force F_B. If the magnetic induction intensity generated by the magnetic slider on the slope is B, and the length of the contact line between the slider and the slope is L, when the slider slides down at a uniform speed v, the electromotive force generated by cutting the magnetic line of force is $E = BLv$. Therefore, the current $I = BLv/R$, in which R is the equivalent resistance induced by magnetic induction. At this time, the reaction force of Ampere force F on the inclined plane is magneto-resistive force F_B, we have

$$G\sin\theta = Kv + \mu G\cos\theta \qquad (6\text{-}3)$$

in which G is the gravity of the sliding block, and θ is the inclination angle between the inclined plane and the horizontal plane, and μ is the sliding friction coefficient between the slider and the inclined plane. Formula (6-3) could be rewritten as

$$\tan\theta = \frac{K}{G} \cdot \frac{v}{\cos\theta} + \mu \qquad (6\text{-}4)$$

113

Obviously, $\tan \theta$ has a linear relationship with $v/\cos \theta$, K/G and μ could be obtained from its slope and intercept.

IV. Contents and steps

(1) Observe the magnetic damping phenomenon and clarify the experimental conditions.

(2) Measure the magnetic damping coefficient and sliding friction coefficient.

Students are required to design experiment steps independently.

V. Data record and process

Students need to design data record sheets by themselves.